SCIENCE ADVENTURERS

AMAZON EXPLORERS

BY ANDREA PELLESCHI

CONTENT CONSULTANT

Nathan Moore
Associate Professor and Graduate Program Director, Geography
Michigan State University

Essential Library
An Imprint of Abdo Publishing
abdobooks.com

ABDOBOOKS.COM

Published by Abdo Publishing, a division of ABDO, PO Box 398166, Minneapolis, Minnesota 55439. Copyright © 2020 by Abdo Consulting Group, Inc. International copyrights reserved in all countries. No part of this book may be reproduced in any form without written permission from the publisher. Essential Library™ is a trademark and logo of Abdo Publishing.

Printed in the United States of America, North Mankato, Minnesota.
092019
012020

 THIS BOOK CONTAINS RECYCLED MATERIALS

Cover Photos: Carl De Souza/AFP/Getty Images, (front); Shutterstock Images, (back)
Interior Photos: iStockphoto, 4–5, 12–13, 24, 30–31, 70–71, 72; Scott Wallace/Hulton Archive/Getty Images, 7; Giacomo d'Orlando/Abaca/Sipa USA/AP Images, 10; Martial Trezzini/Keystone/AP Images, 14; Carlos Mora/Alamy, 16–17; DigitalGlobe/ScapeWare3d/Getty Images, 20–21; Virginie Clavieres/Paris Match Archive/Getty Images, 26–27; Neo Martinez/Science Source, 29; AP Images, 35; Anita Studer/Science Source, 36; Philippe Psaila/Science Source, 38–39; Shutterstock Images, 40–41, 75; GSFC/JPL, MISR Team/NASA, 43; Eric Risberg/AP Images, 46; Ricardo Stuckert/iStockphoto, 48–49; Suamy Beydoun/AGIF/AP Images, 52; Laszlo Mates/iStockphoto, 54–55; Giacomo d'Orlando/Abaca/Sipa USA/AP Images, 57; Bernardo De Niz/Bloomberg/Getty Images, 59; RPBMedia/iStockphoto, 60–61; Encyclopaedia Britannica/Universal Images Group North America LLC/Alamy, 63; Red Line Editorial, 65; Universal Images Group/Getty Images, 68–69; Studio Annika/iStockphoto, 77; Raphael Alves/AFP/Getty Images, 81; Brent Stirton/Reportage Archive/Getty Images, 82–83; Eraldo Peres/AP Images, 86–87; Andre Penner/AP Images, 88–89; Paulo Amorim/Sipa USA/AP Images, 90; Mark Alexander/iStockphoto, 92–93; Wim d'Hooge/iStockphoto, 96–97; Mauro Pimentel/AFP/Getty Images, 98

Editor: Melissa York
Series Designer: Laura Graphenteen

LIBRARY OF CONGRESS CONTROL NUMBER: 2019942078

PUBLISHER'S CATALOGING-IN-PUBLICATION DATA

Names: Pelleschi, Andrea, author.
Title: Amazon explorers / by Andrea Pelleschi
Description: Minneapolis, Minnesota : Abdo Publishing, 2020 | Series: Science adventurers | Includes online resources and index.
Identifiers: ISBN 9781532190315 (lib. bdg.) | ISBN 9781532176166 (ebook)
Subjects: LCSH: Amazonia--Juvenile literature. | Amazon River Region--Juvenile literature. | Scientists--Juvenile literature. | Discovery and exploration--Juvenile literature. | Adventure and adventurers--Juvenile literature.
Classification: DDC 577.4--dc23

CONTENTS

CHAPTER ONE
UNCONTACTED PEOPLE 4

CHAPTER TWO
MYSTERIOUS EARTHWORKS 16

CHAPTER THREE
BRAZIL'S HIGHEST MOUNTAIN 30

CHAPTER FOUR
THE AMAZON'S CORAL REEF 40

CHAPTER FIVE
MEDICINE DISCOVERIES 48

CHAPTER SIX
STRANGE WEATHER 60

CHAPTER SEVEN
NEW SPECIES 70

CHAPTER EIGHT
DEFORESTATION 82

CHAPTER NINE
CLIMATE CHANGE 92

ESSENTIAL FACTS 100
GLOSSARY 102
ADDITIONAL RESOURCES 104
SOURCE NOTES 106
INDEX 110
ABOUT THE AUTHOR 112

CHAPTER ONE

UNCONTACTED PEOPLE

In 2002, an expedition traveled deep into the Amazon rain forest. The travelers weren't on a trail, so they bushwhacked through the underbrush, slogged across mud, and ducked under branches. Nearby, a small river meandered into and out of view, one of the Amazon's tributaries. Every now and then the expedition members glimpsed its rushing water through the green brush.

Their mission was to find the Flecheiros, or "Arrow People." The Flecheiros were an uncontacted tribe, one of about 80 of such tribes.[1] There are about 400 indigenous tribes in the Amazon in total.[2] Uncontacted tribes do not interact with the dominant society in their area, such as government officials, scientists, loggers, miners, or others. However, they may interact with other tribes, or they may have had contact with the dominant society in the past but no longer wish to do so.

Many people of the Amazon mix traditional culture with influences from outsiders.

Many days before, the group of more than 30 men had sailed up the Itaquai River. When the water became too shallow, they left their boats and set off on foot to the floodplain of another river, the Jutaí. Months before, the leader of the expedition, Sydney Possuelo, had mapped the Flecheiros' whereabouts. He'd located their malocas, or longhouses, from the air and marked their locations with GPS. Now he used a compass to guide the group through the forest.

NO-CONTACT POLICY

The goal of this expedition wasn't to make contact. It was to find the Flecheiros, discover how well they were doing, map their territory for protection, and leave without disturbing them. Possuelo used to be a *sertanista*, or Brazilian explorer. In the early 1900s, the sertanistas' mission was to discover tribes and introduce them to modern life so settlers and loggers could more easily move into the area. Often sertanistas captured tribe members and forced them to work as manual laborers in agriculture and mining. Sertanistas also brought disease and illness, which decimated tribes.

THE AMAZON

The Amazon rain forest in South America is the largest on Earth. More species live in the rain forest than in any other ecosystem on the planet. This rich biodiversity includes 40,000 plant species, 1,300 bird species, and 430 mammal species.[3] Scientists also estimate there are about 2.5 million species of insects.[4] In the heart of the rain forest flows the Amazon River. It lies just south of the equator and is the second-longest river in the world, behind the Nile in Africa. The Amazon river basin—the area where water drains into the Amazon, including its 1,100 tributaries—is about the size of the continental United States.[5] The lowland areas near the Amazon and its main tributaries flood regularly and are called floodplains. Most of the basin is at a higher elevation, or upland, and is called *terra firme*. Rain forest covers most of the basin, but the higher northern and southern edges have dry forest and savanna. The Andes region to the west has montane forest with large trees such as pines.

Sydney Possuelo, *wearing hat*, **with a group of scouts from the Matis people, 2002**

By the 1950s, many indigenous people had died, mostly from disease. Anthropologist Darcy Ribeiro believed they would all be gone by 1980. Then, in 1967, the numerous crimes against indigenous people were listed in a report, including murder, enslavement, and land theft. The report became famous and led to changes in the Brazilian government. Eventually, the indigenous population started to grow. Today, outsiders who work with these tribes are more often called *indigenistas* or ethnographers.

LAST REMAINING MAN

In the forests of Rondônia State in Brazil lives a man who is the last of his tribe. Anthropologists believe ranchers killed off the rest of the tribe and the man has been living alone ever since. FUNAI has been monitoring him since the 1990s by tracking his footprints, chopped trees, and traps for prey. In 2018, they released a 2011 recording of him cutting a tree with an axe. In the video, he appeared to be in good health and in his late 50s. Officials have tried to contact him, but he has shown no interest, even firing an arrow at an official in 2005. In May 2018, footprints and a cut tree were seen, indicating that he was still alive.

In the 1980s, Possuelo advocated for leaving tribes isolated in their own environments. In 1987, he created a unit within Brazil's National Indian Foundation (FUNAI). The unit's mission was to investigate reports of uncontacted indigenous people, verify that they are thriving, and do what is necessary to make sure they are left alone. One way of doing this involved the creation of reserved lands that no one from the outside could enter and harm through deforestation, mining, logging, or agriculture.

The Flecheiros lived in a protected territory called the Javari Valley Indigenous Land. This was one of six special regions that Brazil had set aside to protect both indigenous cultures and the millions of acres of biodiversity in which the people lived.

FINDING THE FLECHEIROS

Usually expeditions to find uncontacted tribes went smoothly. But on this rare occasion, the expedition ran into trouble. The Flecheiros were known for being excellent archers and hostile to outsiders. As the expeditioners cut through the brush, they discovered a path, straight and well used. Possuelo led the group onto it. They marched cautiously forward. Birds cried out all around them.

"That's not a bird," said Possuelo. "They're signaling to each other. They're watching us."[6]

The group kept walking. They saw footprints in the red dirt, pointed in the direction they were going. Suddenly, they came upon a sapling bent across the path like a gate. It was clear someone had deliberately wrestled the small tree into a horizontal position, about four feet (1.2 m) off the ground.

Possuelo knew what that meant. "Stay out. Go no farther." He figured they were close to the Flecheiros' village, and he ordered everyone off the path. That's when they noticed two of their expedition members were missing. Possuelo and several others went searching. They discovered a village farther up the path with several large huts. The village was empty, and it looked like the villagers had left in a hurry, probably frightened by the intruders. Five or six fires were still smoldering. Smoked meat was piled nearby: crocodile, sloth, monkey, tapir, and more. A couple of ceremonial masks made of bark lay near the fire. Inside one of the huts was a broken blowgun and a pot of curare. This was a poison the Flecheiros put on the tips of their arrows and on darts used in blowguns.

The missing men were not in the village, but they turned up later, unharmed, having hidden from the Flecheiros in the brush. In the end, Possuelo considered the mission a success. They'd learned the

Curare is a poison made from plants that causes muscle relaxation and paralysis.

The majority of Amazon tribes have contact with outsiders, and some support themselves by making handicrafts to sell to tourists.

Flecheiros hunted and grew food. They made masks, clay pots, arrows, and blowguns. They were living independently and seemed to be living well.

WHAT THE SCIENTISTS THINK

The Amazon rain forest touches nine countries, and most of them follow Brazil's example of leaving tribes in isolation. Anthropologists believe this is the best way to protect them. However, isolation isn't always practical or realistic.

One reason is that there really is no such thing as a completely isolated tribe. Most of them have seen planes in the sky, had conflict with loggers, or seen other societies. In fact, isolated tribes are often interested in contacting other people. Anthropologists such as Kim Hill from Arizona State University have interviewed people who left isolated areas. They found that fear was what kept them from making contact, not lack of interest. "There is no such thing as a group that remains in isolation because they think it's cool to not have contact with anyone else on the planet," said Hill.[7]

In addition, oil and gas companies, loggers, and other developers are encroaching on reserved land. Many isolated tribes see tourists, Christian missionaries, and villagers. Hill believes that it is better to contact indigenous people before they have a violent clash with outsiders or contract a disease for which they have no immunity. He thinks anthropologists should slowly get to know the tribes and build a relationship with them. Then the anthropologists should live with the tribes and monitor them. If an illness broke out, they could respond to it rapidly before the entire tribe

HISTORY OF ANTHROPOLOGY

Anthropology is the science of human beings and their ancestors, and it includes social relations, physical characteristics, environments, and how cultures develop. In the 1700s, modern anthropology began with the Enlightenment. This was a time when Europeans began to use reason as a way to advance science and knowledge. They applied this principle to human behavior and societies, particularly the indigenous people living in their colonies, whom they considered inferior. In the 1800s, anthropologists believed all cultures could be compared with each other, and certain ones were superior to others. By the 1900s, however, anthropologists strove to understand other cultures on their own terms and not compare them with each other. Some anthropologists became participant-observers and lived within different cultures for extended periods of time. Others began studying cultures in the developed world rather than focusing on the developing world. Today, anthropologists use new methods to help their understanding, such as CT scans, genetic tests, and digital technology.

became sick. But there is no standard policy across all Amazon countries. Each country has its own policy, and often the interests of developers outweigh the interests of indigenous peoples.

In Brazil, the situation was changing rapidly at the end of the 2010s. After being elected in October 2018, President Jair Bolsonaro planned to open up more land to commercial development. He changed the agency that is in charge of reservation boundaries. The Agriculture Ministry took over this responsibility from FUNAI. And the Agriculture Ministry's mission is to promote, expand, and improve agriculture, which often conflicts with protecting indigenous people.

This change emboldened developers to illegally claim land set aside for indigenous people. In March 2019, members of the Uru-Eu-Wau-Wau tribe had a standoff with developers who came to the border with chain saws, machetes, and guns. The Uru-Eu-Wau-Waus confronted them with a mix of

Deforestation in the Amazon threatens its people and animals, as well as the health of the entire planet.

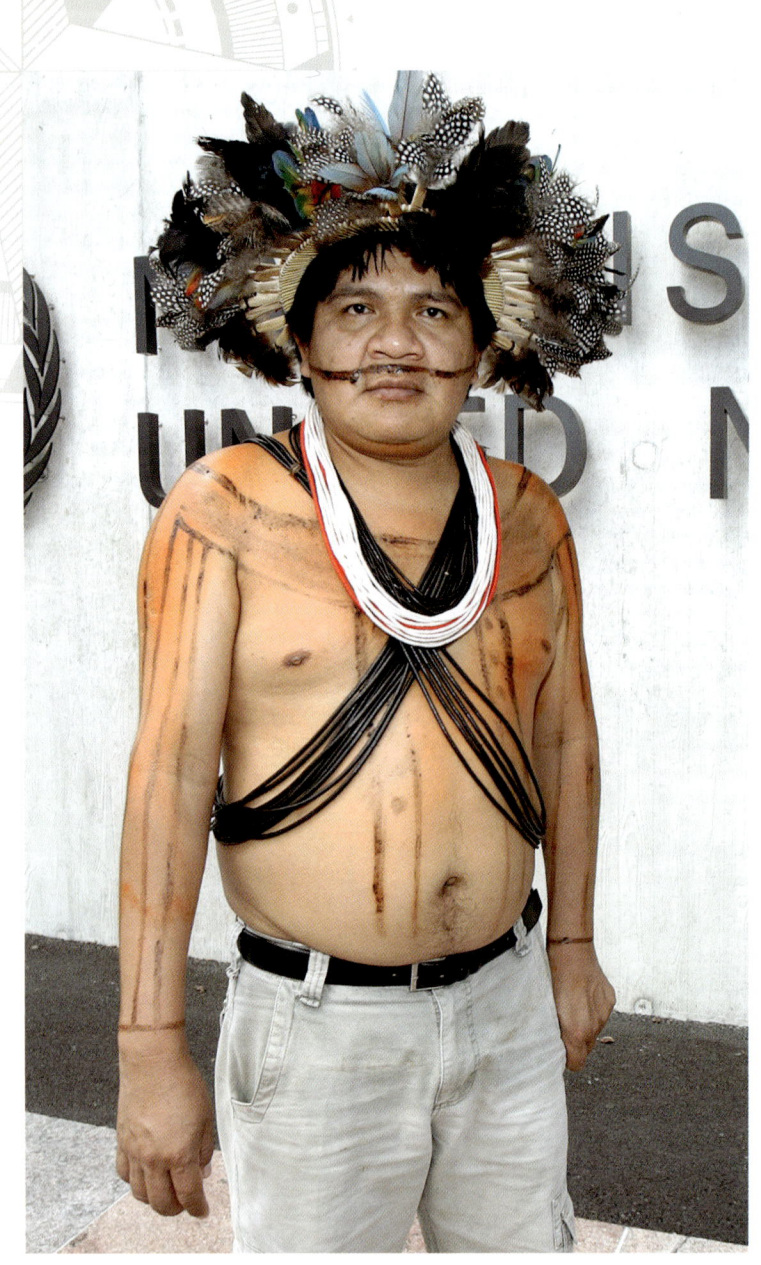

Almir Narayamoga Surui, a leader of the Paiter-Surui tribe, visited the United Nations in Switzerland to speak out against deforestation in the Amazon.

old and new technology: bows and arrows dipped in poison and cell phones and computers, which they used to record the confrontation. The developers retreated, but this was just one of many threats that have occurred under President Bolsonaro.

In Brazil, FUNAI and nongovernmental organizations such as Instituto Socioambiental (ISA) are working to protect indigenous peoples. FUNAI's General Coordination Unit of Uncontacted Indians—begun by Possuelo—is dedicated to those who still live in isolation. In addition, most tribes are working to save themselves, including the Paiter-Surui tribe, which uses technology to protect itself and its land. The people who live in the forest are the best forest managers.

SCIENCE IN THE AMAZON

Learning about uncontacted tribes is just one of many scientific missions in the Amazon. And every scientist who works in the Amazon must be aware of indigenous cultures and indigenous peoples' rights. Because it is the largest tropical rain forest in the world, scientists live there or travel there to explore its land, rivers, plants, and animals. They bring submersibles and machetes and computers. They climb sky-high observation towers or ride helicopters through the clouds to visit Brazil's highest peak. They study satellite images, collect specimens, and make discoveries that inform the world about the wonders of the Amazon.

USING TECHNOLOGY

Chief Almir Narayamoga Surui is head of a clan in the Paiter-Surui tribe. He uses Google Earth to track illegal forestry in Brazil's state of Rondônia. These images show blots of brown earth amid the darker green of the forest. Using geographic software, he makes maps of the areas and pinpoints areas of greater risk. The chief also uses smartphones to prevent his tribe members from overhunting. He hopes this technology will help him keep the 1,000 square miles (2,590 sq km) where the tribe lives from being cut down illegally. It is one of the few remaining forested areas in Rondônia.

CHAPTER TWO

MYSTERIOUS EARTHWORKS

It is a common misconception among the general public that the rain forest was sparsely populated before Europeans conquered South America. Many believed that hunter-gatherers lived a nomadic life in the rain forest and made little impact on their environment. People thought low populations were normal, not understanding the dramatic impact European diseases had in reducing the South American population. However, scientists have known this wasn't true for decades, and new evidence backs this up.

Archaeologists unearth funerary urns left by the Marajoara people from Marajó Island near the mouth of the Amazon River. The people are also known for their large earthworks and mounds.

EXISTING EARTHWORKS

Before 2018, scientists knew there were pre-Columbian societies along the southern rim of the Amazon. Pre-Columbian refers to the time before Christopher Columbus came to the Americas in 1492. Scientists had discovered evidence of these societies in the form of earthworks—artificial ridges or mounds formed into shapes. Some of the earthworks were circles, squares, or hexagons. These are called geoglyphs. Others showed outlines for villages, fortified settlements, canals, and other structures.

The earthworks stretched from the Brazilian state of Acre to the savannas of Bolivia to the forests of Upper Xingu, another region of Brazil. Many scientists believed these earthworks indicated that the pre-Columbian Amazon rain forest had been heavily populated. They believed there could be many more undiscovered earthworks and that an uninterrupted swath of them could stretch 1,100 miles (1,800 km) east to west in the southern Amazon.[1]

There was just one problem with this theory. A particular area of the southern rim of the Amazon had not been explored, so there was no evidence that the features extended

GEOGLYPHS

Ancient peoples have created geoglyphs all over the world. Among the most famous are the Nazca lines in western Peru. Etched into the desert are more than 1,000 straight lines, geometric figures, and shapes of animals and plants. They were made by the Nazca people sometime between 1 and 700 CE. In Britain, an image of a horse is carved in a chalk hill, created between 1740 and 210 BCE. In Kazakhstan, about 260 geometric shapes were created by building up dirt, rocks, and other material. Experts are unsure of their age, theorizing they are either about 2,800 or 8,000 years old.

as far as scientists suspected. This area was the upper Tapajós river basin. The Tapajós is a tributary of the Amazon.

The Tapajós river basin is about three times the size of Florida.

NEW EARTHWORKS

In 2018, a study was published in *Nature Communications* about a recent discovery of new earthworks in the upper Tapajós basin. Jonas Gregorio de Souza, archaeologist at the University of Exeter, was the lead author of the study. He and other researchers used satellite imagery to explore this area. They were able to do so because of deforestation. What had once been obscured by trees was now out in the open.

The satellite images showed geoglyphs that were round or square, possibly used for ceremonial purposes, many longer than football fields. Some looked like giant hoops of upraised brown earth. Others looked like tan or reddish gouges in green farm fields. Some of the earthworks were large and had nothing else nearby. Others had a combination of small and large earthworks. Researchers found ditches, walls, roads, fortification sites, and platforms for houses. Overall, they discovered 81 sites.[2]

SATELLITE IMAGERY

As satellites orbit Earth, they use sensors to obtain information about weather, roads, forests, and many other things. The sensors measure different types of electromagnetic radiation, or light. This includes visible light, infrared light, ultraviolet light, and more. First, the sun shines onto Earth. Some light is reflected off concrete, trees, clouds, and other formations back to the satellite, and an image is created. Because the sun cannot emit all types of electromagnetic radiation, some satellites emit their own, giving scientists additional data. Images created from visible light look like photographs. Other images might have vibrant green or red colors to represent light we cannot see, such as infrared light.

After studying the satellite images, the research team picked 24 to study in person.[3] They trekked into the field and visited each site, pleased to discover the satellite imagery was correct. They found the traces of people who had lived in the area long ago. Many sites were near springs and small streams, which contradicted previous theories that Amazon people lived only near large rivers or on floodplains.

At the earthworks, researchers often walked along large ditches under the hot sun, following what might have been defensive barriers for villages long ago. At one of the sites, they found pieces of pottery, charcoal, and polished axes. They also found dark soil, which is an indication that people had farmed the land and composted with charcoal.

When the charcoal was carbon dated, it was found to be from around 1410 to 1460 CE. This falls

Deforestation and new satellite technology have revealed many earthworks, seen as white lines, that were long hidden by the forest.

within the time period of the previously discovered earthworks along the southern rim, from 1250 to 1500.

"The idea that the Amazon was a pristine forest, untouched by humans, home to scattered nomadic populations . . . we already knew that was not true," said de Souza. "The big debate is how populations were distributed in pre-Columbian times in the Amazon."[4]

To find that out, de Souza and the other researchers went back to the office and created a computer model. They used measurements of soil pH, precipitation, and elevation to map out where additional earthwork sites might be located.

The model showed that the pre-Columbian people lived much closer to each other than previously thought, and they lived in large groups, not small ones. The model indicated that as many as 1,300 earthworks might be found in an area of 154,000 square miles (399,000 sq km) in the southern rim. They estimated that there could be more than 1,000 villages.[5] In addition, the model said that 500,000 to one million people lived in just 7 percent of the Amazon. Scientists used to think that number was much lower—about two million people in the entire Amazon basin.[6] This study shows that, before Europeans arrived, much of the Amazon rain forest

DARK EARTH

Terra preta, Portuguese for "dark earth," is a type of soil that is almost black in color. It is made by slashing and burning vegetation or by smoldering organic matter, such as bones or straw. Pre-Columbian people might have used the slash-and-burn technique to make space for villages. Unlike typical Amazon soil, which is orange-yellow and not very useful for planting, dark earth is extremely fertile. It is often found near archaeological sites. Its presence is a sign that pre-Columbian people lived in the area.

was densely populated with complex societies. Then European conquest and disease killed off most of the indigenous people.

PLANTS

Native Amazon people didn't just build earthworks and villages. They also cultivated plants and trees. In 2013, ecologist Hans ter Steege and other researchers went to the Amazon. They looked at 1,170 plots of land that were in remote areas of the forest, far from modern cities and villages. They inventoried all the trees in each plot—more than 500,000 of them—and found more than 16,000 species. However, there was very little variety in about half of the plants they inventoried. About half of the trees they counted were made up of just 227 species, or just over 1 percent of the total species they cataloged. And about 20 of these species consisted of domesticated trees, such as the Brazil nut and the ice cream bean tree.[7]

Why were there so many similar plants? Steege theorized that pre-Columbian people might have sowed these plants and trees on purpose. He teamed up with archaeologists to study how close to sites of ancient villages the domesticated plants could be found. They

EUROPEAN CONQUEST

When the Europeans first came to the Americas in 1492, it is estimated that 60 million indigenous people lived there, and they farmed about 10 percent of the land. However, the Europeans brought diseases such as smallpox, measles, and influenza. They also brought warfare and famine, and they enslaved the indigenous people. By the start of the 1600s, about 90 percent of the indigenous population had died.[8] Fields returned to jungle. As this happened, the amount of carbon dioxide in the atmosphere decreased. Plants take in carbon dioxide through photosynthesis, so the more plants and trees, the less carbon dioxide. Carbon dioxide helps keep the atmosphere heated. A 2015 study in *Nature* argues that this reforestation caused the entire planet to enter a period of global cooling in the 1600s.

The ice cream bean tree grows beans in long pods.

found that more of these plants were found closer to the villages. Often the types of domesticated plants were not local. People must have brought them to their villages from far away.

A 2017 study in the journal *Science* followed. It looked at archaeological sites, environmental information, and the distribution of plants near pre-Columbian sites. Study authors, led by Carolina Levis from Wageningen University, found that 85 species of trees were more likely to be near places where humans lived at that time. And these 85 trees weren't just random wild trees found in the area.[9] They were domesticated. Like the 2013 study, researchers concluded that greater numbers of these domestic trees were found as they got closer to the sites.

According to the study, pre-Columbian people had a much larger impact on the

Amazon rain forest than was originally thought. They created gardens and cultivated plants for more than 8,000 years. They chose trees they liked and planted them near their villages. Today, there are still areas with large populations of trees planted by pre-Columbian people, especially in the southwest. These include rubber trees, brazil nuts, cocoa trees, and maripa palms. "People arrived in the Amazon at least 10,000 years ago, and they started to use the species that were there," said Levis. "They really cultivated and planted these species in their home gardens, in the forests they were managing."[10]

About 390 billion trees grow in the Amazon rain forest.[11]

Today, experts disagree on what the plants mean. At least 500 years have elapsed since pre-Columbian people lived in the Amazon. A lot could change in that time. Plants and trees might be growing in certain areas for reasons that have nothing to do with humans. For instance, there used to be a theory that the Maya peoples grew their own breadnut trees because so many were found near Mayan ruins. It turns out, though, that bats carry the breadnut seeds and drop them all around. Plus, Mayan ruins have limestone. The breadnut trees may have grown near them for the minerals in limestone, not because Mayas planted them there.

Still, the 2017 study was the most comprehensive of its kind. It used information from a database of archaeological sites and from a catalog of Amazon animal and plant species. The study correlated 80 years of research on the relationship between living things and their environment, as well as the indigenous people who lived there. At archaeological sites, the study looked at rock paintings, earthworks, mounds, ceramics, and dark earth. Anything that would give researchers clues as to how humans influenced the Amazon long ago was considered. However, the study makes clear that more research needs to be done. For instance, scientists could study whether mounds, ceramics, or dark soil might indicate which trees were planted in particular areas.

Archaeologists study the rock paintings some Amazon cultures created.

Computer models can help researchers understand everything from what atoms are made of to how galaxies interact. This model shows food webs in a rain forest ecosystem.

SCIENCE CONNECTION
MAKING A COMPUTER MODEL

Computer models use mathematics to create a simplified version of real life. They can be used to study something that happened in the past or to predict something that may happen in the future. These include things like weather, the stock market, the movement of molecules in a balloon, and the number of pre-Columbian earthworks in the Amazon.

Scientists use information from the past to predict the future. For instance, they may look at weather in the past to predict future weather. In the case of pre-Columbian earthworks, scientists look at previously discovered features, plus environmental and terrain data, to predict other features.

Scientists often have to make assumptions about their model and simplify their data. Sometimes they have to average some data or not include other data. Then they run their model to see whether it works. They test it by inputting historical data and seeing if the results match what actually happened. If so, scientists can use the model to help forecast future events.

In the case of pre-Columbian earthworks, scientists ran their model with all of the data. Then they ran the model with a filtered set of data to eliminate sampling bias. Bias might come from using too much data from areas with a large number of earthworks or by counting earthworks as multiple sites when they are actually just one site. Then scientists compared the initial results with the filtered results, and they found the outcomes were nearly identical. After further testing to eliminate errors, they determined that their model was accurate.

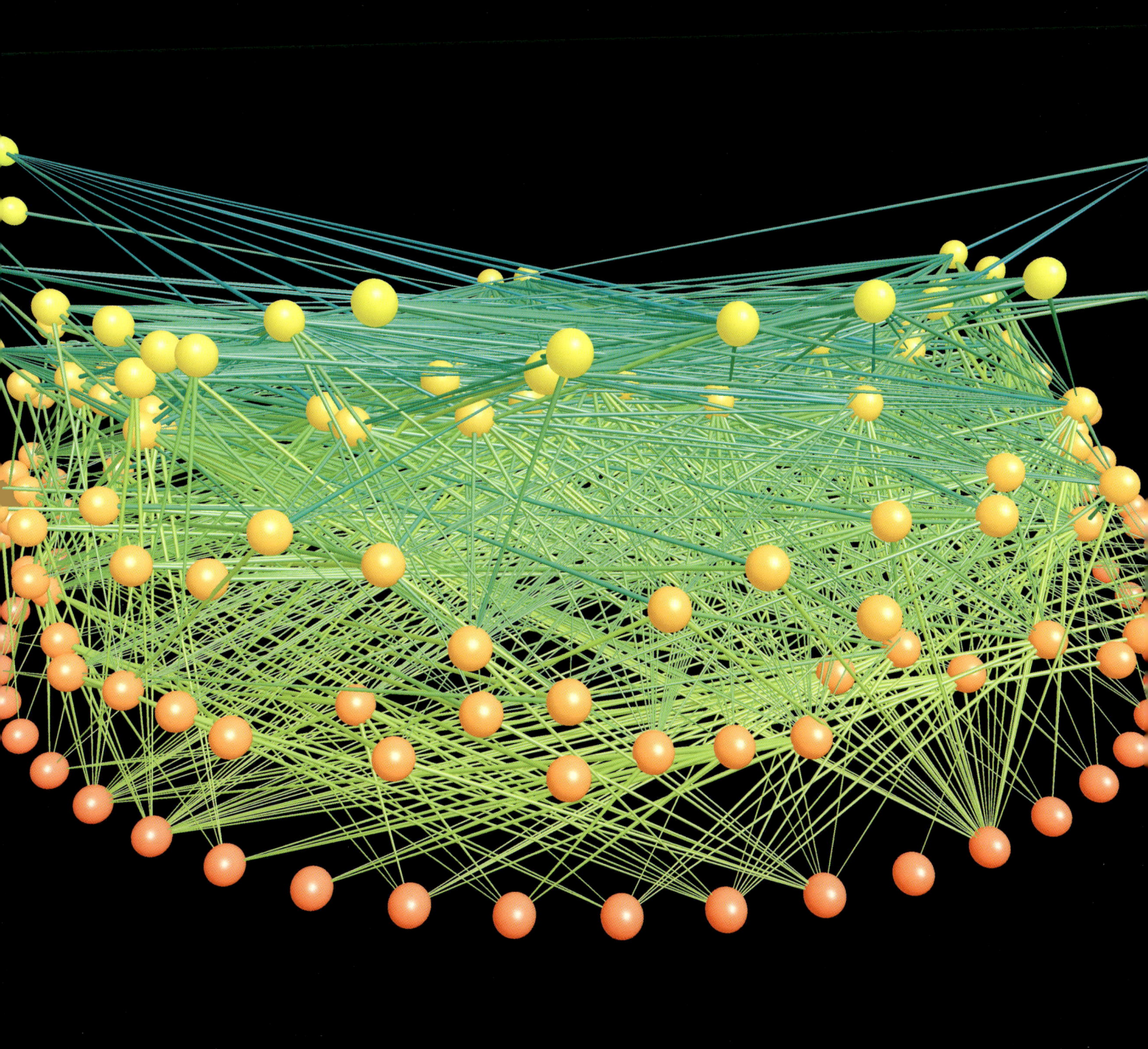

CHAPTER THREE

BRAZIL'S HIGHEST MOUNTAIN

As the helicopter took off from the cleared field of the army base, the dense Amazon rain forest fell away, growing smaller and smaller. On board were biologists and professors from the University of São Paulo. They were going to explore Pico da Neblina, Brazil's highest mountain. The border between Brazil and Venezuela runs through Pico da Neblina. Their mission was to find new species of plants and animals, understand how these species came to live on the mountain, and see how climate change is affecting them. The scientists were some of the first people to explore the mountain in this way. It was a rare and exciting opportunity.

Pico da Neblina is often shrouded in clouds.

GETTING THERE

In 2003, the Yanomami people and Brazilian government closed the state park surrounding Pico da Neblina. Illegal mining and excessive tourism were damaging the protected area. But in 2017, they granted permission to the scientists to enter the state park and travel to the mountain.

One reason Pico da Neblina hadn't been explored by scientists before is that it is in a remote area of the northwest Amazon rain forest. To get there, the research team had to fly from São Paolo to Manaus, travel up the Rio Negro for three days, drive 50 miles (80 km) to another river, and take another boat ride. Eventually the researchers made it to a small army base where they stayed until the final leg of the journey, the helicopter ride.

Pico da Neblina means "Mountain of Mist." That day, as the helicopter neared the mountain, they saw it was named appropriately. Clouds circled the mountain, hiding it from view. As the helicopter passed through the mists, the scientists caught their first glimpse of the

ILLEGAL MINING

In December 2018, a study revealed a new increase in illegal mining in the Amazon rain forest. A group of nongovernmental and environmental organizations mapped more than 2,300 mining sites across six countries.[1] Called *garimpo*, or artisanal mining, this mining takes place throughout the rain forest, including in nature reserves and protected indigenous areas. It causes devastating effects on the environment: forests are cut down and ponds are dug into the earth. Miners use mercury to pull gold out of rocks, and the toxic substance leaks into rivers, poisoning the water. Indigenous people believe that they have little choice but to work in the mining industry if they want to survive. As of 2019, the government of Brazil wanted to open up more protected land for mining, which could further endanger these areas.

mountain, which looked like a gray stone fortress with steep jagged walls jutting up from a sparse field.

The Pico da Neblina environment is very different from the rest of the rain forest. The peak reaches about 10,000 feet (3,000 m) above sea level. At that high elevation, even near the equator, the air is cool enough to make people don jackets. Instead of thick tree cover, low-level vegetation and rocks cover the ground, similar to a savanna. "Visiting this place is like taking a step 1,000 years ago," said Miguel Trefaut Rodrigues, lead scientist of the expedition and professor at the University of São Paulo.[2] Rodrigues is a reptile expert who has discovered more than 80 new species.[3] As he further described it, "We know the landscapes found at altitudes above 1,700 m [5,600 feet] have nothing to do with today's Amazonia. They consist of open types of vegetation, such as high-altitude meadows and grasslands, with a much cooler climate than the tropical forest."[4]

EXPLORING

Since they only had provisions for ten days, the scientists went looking for rare species right away. With Yanomami guides, they hacked through the low brush and began exploring. Almost by accident, they came upon a tree frog. It was bright orange with long fingers for grasping branches. Its iridescent eyes shone with many colors. One of the scientists theorized that the frog's eyes and skin had adapted to being at a higher elevation where there was more radiation from the sun. Was it a new species? Possibly. They would take it back to the laboratory for study, to see whether the frog had already been discovered.

THE YANOMAMI

With a population of about 32,000, the Yanomami people constitute the largest isolated tribe in South America.[5] They live in a protected area in southern Venezuela and northern Brazil. They often dwell in a round enclosure called a *yanos* or *shabonos* that can hold 400 families at a time. Inside the enclosure, individual families have their own hearths where they prepare food during the daytime. At night, the Yanomami string up hammocks and make fires to keep themselves warm. Both men and women fish. The men hunt monkeys, deer, fowl, and tapir, while the women plant gardens and harvest cassava, tubers, corn, and more. Gardening accounts for about 80 percent of their diet, and the Yanomami use about 500 different plants for medicine, food, and building materials.[6]

Another scientist found a brown-and-tan lizard with mottled skin. Like the frog, it would be taken to the lab to see whether it was a new species. The researchers used models to see how creatures such as the lizard would cope with climate change. Lizards cannot regulate their own body temperatures. They match the temperature in their environment. So on a hot day, they stay in the shade to cool down. But what if the temperature is still too hot in the shade because of climate change? Creatures like the lizard might die out.

NEW SPECIES

One of the goals of the mission was to understand how certain species came to live on Pico da Neblina. Such was the case with a new species of lizard nicknamed Night Sky. It was called this because of the black and white spots on its body that looked like stars. Night Sky is closely related to animals in the *Riolama* group. These animals live in this part of South America, but not on Pico da Neblina.

How did Night Sky come to live on the mountain, separate from other related species? One theory is that Pico da Neblina was once part of a large plateau. Over the course of millions of years, the plateau eroded until only mountains remained. "The hills we see today are just the survivors, the relics of a much higher plateau in the past," said Ivan Prates, a researcher from the Smithsonian National Museum of Natural History.[7] Specimens like Night Sky will undergo DNA testing to see where they fall in their family tree. Then that information will be checked against geological information.

Besides Night Sky, the research team discovered two additional species that had become isolated from other similar species: Chubby Grey and Lizard-Walking Toad. Chubby Grey was discovered on the grass in one of the scientists' tents. A warty, dark-brown frog, it is part of the *Terrarana* family. Lizard-Walking Toad was found sleeping on a leaf. It walks like a lizard instead of hopping, and it is part of a group that lives in Venezuela, Guyana, and northern Brazil. Like Night Sky, both Chubby Grey and Lizard-Walking Toad might help

New species are discovered frequently in the Amazon. The black-headed sagui dwarf monkey was discovered in 1997 and is only four inches (10 cm) tall.

Multiple species of pygmy owls live in Brazil.

explain how Pico da Neblina was connected to other highland areas in the past and how those connections have since eroded away.

Several other new species were discovered on the mountain. Big-Eyed Red is a frog with smooth reddish-brown skin. It was found on the top of the summit, sitting under a rock. Chirping Frog was found in the foothills. It makes a call like a bird and belongs to the *Allobates* group. It is a type of nurse frog, whose males carry eggs on their backs. Plump Digger is an unusual-looking frog with plump legs and a pointy nose, which it uses to burrow into the earth. It is about one inch (2.5 cm) long and makes a distinctive call. "They come to the surface during humid nights and go for a walk. That's how we got them," said Prates.[8]

The researchers found one new species of owl: the Neblina pygmy owl. It has gray feathers, yellow feet, and a distinct singing voice. It was discovered in a tree about five feet (1.5 m) off the ground. Before it was captured, the researcher made sure to record its song.

Scientists nickname their finds because it can take several years for a new species to receive an official scientific name.

NAMING A NEW SPECIES

With more than eight million species worldwide, it's important to come up with a unique name for each one.[9] The International Code of Zoological Nomenclature gives instructions on how to do this. First, scientists must determine the genus name, which describes the category of related species. Then scientists must come up with a unique species name. This should describe the habitat, behavior, location, or appearance of the new species. Scientists can be creative. For instance, the *Peckoltia greedoi* catfish from Brazil was named after the Star Wars character Greedo because its discoverers saw a strong resemblance.

A new lizard named Brown Giant was discovered in the kitchen by a cook. It was about five inches (13 cm) in length. Before it was captured, it lost its tail. This is a typical way lizards try to avoid being captured. Finally, a new species of plant was found. Called Neblina Phyllanthus, this shrub has small green teardrop-shaped leaves and is about six and one-half feet (2 m) tall.

Overall, this expedition found nine new species, including frogs, lizards, an owl, and a plant. Researchers also studied the area around the peak and collected more than 700 specimens in total.[10] The entire expedition from start to finish took 30 days. From there, the specimens were transported out of the area to be studied at the university.

Amazon explorers go to great lengths—and heights— to study their subjects.

CHAPTER FOUR

THE AMAZON'S CORAL REEF

The Amazon River picks up 1.3 million short tons (1.2 million metric tons) of sediment every day as it travels from Peru in the west to the Brazilian coast in the east.[1] By the time it spills into the Atlantic Ocean, the water is brown and dense. Scientists suspected a coral reef lay under that muddy water. A 1977 report documented a type of fish that is typically found near coral reefs. But the plume of a large river, where fresh water meets the salty water of an ocean, does not have a typical environment for a healthy coral reef. The salt content, acidity, and low levels of sunlight there are not ideal, so many people doubted the existence of a reef. However, a 2012 scientific expedition discovered the surprising truth.

The Amazon River is murky and muddy in many places.

The Amazon River has more sediment than any other river in the world.

CORAL

Coral is a type of invertebrate, an animal without a spine. It is made up of thousands of small polyps, which have a mouth and a stomach. There are soft corals that live in warm water but do not create reefs, stony corals that live in warm water and build reefs, and deepwater corals that like dark, cold water. Most reef-building coral and some soft coral have algae inside their tissue. The algae, called zooxanthellae, carry out photosynthesis. This gives the coral oxygen and vital nutrients. In turn, the coral provides a safe environment for the algae as well as the material it needs for photosynthesis.

DISCOVERY

In 2012, Patricia Yager, an associate professor from the University of Georgia, wanted to study the Amazon's plume. Specifically, she wanted to see how it was affecting the way the Atlantic Ocean was absorbing carbon dioxide. But in order to take a ship to the mouth of the Amazon, she needed the Brazilian government's approval. Officials requested that she bring Brazilian oceanographers on the ship. One of the oceanographers was Rodrigo Moura. He had a paper from 1977 with a hand-drawn map of a possible coral reef under the plume. "I kind of chuckled when Rodrigo [Moura] first approached me about looking for reefs. I mean, it's kind of dark, it's muddy—it's the Amazon River," said Yager.[2] But she was willing to look for reefs.

Along with the team of scientists, Yager sailed the research vessel *Atlantis* to the plume. While she conducted her study on carbon dioxide, Moura watched the sonar for signs of a coral reef on the seafloor. He mapped the area that looked promising. Then Yager released metal dredges and trawls to collect samples. The dredges brought up corals, sponges, fish, and sea stars.

After this initial discovery, Moura and other researchers came back to the reef and took more samples in 2014. They discovered that the plume itself varies. It covers the southern part of the reef only three months out of the year. This allows more photosynthesis to occur and more colorful coral to grow. The northern part of the reef is covered by the plume about half of the year, so it has mostly sponges and fish.

Before this reef was discovered, scientists used to think there was a gap between a coral reef up north in the Caribbean Sea and a reef down south along the South American coast. The Amazon River bridges the gap. It also shows how corals can survive in harsher environments, which gives hints on how corals will endure climate change. "Tropical coral reefs are in decline worldwide," said Rebecca Albright, an oceanographer and coral researcher at

As shown by satellite image, the Amazon's plume reaches far into the ocean.

43

the Carnegie Institution for Science. "It may become more important to understand which organisms can tolerate harsher conditions."[3]

SPONGES

Sponges live within coral reef ecosystems. Like corals, they are invertebrates that do not move. They can live in salt water like coral, but they can also live in fresh water and estuaries, where rivers flow into oceans. These simple organisms have a strong, porous structure through which they filter water. This helps protect the coral reef from changes in temperature, light, and nutrients. Sponges come in many colors, perhaps to protect them from the sun's ultraviolet rays. They can live on hard surfaces as well as mud and sand, but they almost never float free. There are about 8,550 species of sponges in the world.[8]

WHAT THEY FOUND

The results of the expeditions to the coral reef were published in the journal *Science* in 2016. All told, the researchers discovered a coral and sponge reef that was 600 miles (970 km) long, taking up an area of 3,600 square miles (9,300 sq km) on the continental shelf.[4] The reef stretched from French Guiana to the state of Maranhão in Brazil and extended up to 75 miles (120 km) offshore. Despite the low light and little oxygen, some of the coral structures grew up to 100 feet (30 m) high and nearly 1,000 feet (300 m) long.[5]

Researchers also found an abundance of life: 73 types of reef fish (mostly carnivorous), 61 types of sponges, 35 algae, 26 soft corals, and 12 stony corals.[6] Plus they found basket stars, brittle stars, segmented marine worms, rhodoliths (a type of pink algae, resembling coral), sea fans, and more. "We brought up the most amazing animals I've ever seen on an expedition like this," said Yager. "All the scientists just hung over the rails amazed at what we were finding."[7]

VIDEO SURVEY

When the reef was first discovered, scientists weren't able to actually see it. The water was too dark, the sea was too rough, and the currents were too dangerous. So they relied on sonar to map the area while dredging to obtain samples. But in 2017 and 2018, new expeditions set out to take the first-ever video surveys of the reef. The results were published in the journal *Frontiers in Marine Science* in 2018.

The surveys were conducted by scientists from Brazilian universities and the environmental organization Greenpeace. They set out in Greenpeace's ship *Esperanza* and sailed to the reef. There, they launched a DeepWorker submersible. This is a small submarine used for scientific exploration. Researchers collected samples off the coast of Amapá and Pará in Brazil. They took photographs and video from the submersible as well as from an underwater camera that could be dropped or towed from a boat.

A giant sponge found in the coral reef weighed as much as a baby elephant.

DEEPWORKER SUBMERSIBLE

The DeepWorker submersible is a small, lightweight vessel used for deep ocean exploration. One person at a time can descend to a depth of 3,300 feet (1,000 m) for up to 80 hours.[9] A touch screen computer controls lights, monitors depth, and performs other tasks. Steering is done by foot pedals. The right pedal moves the vessel forward or backward. The left makes it go left or right. Manipulator arms allow the scientist to cut cables, grab and hold objects, and use instruments in the water.

The DeepWorker submersible is not much larger than its passenger.

They discovered that the reef may be up to six times larger than originally thought, with an area of 22,000 square miles (57,000 sq km). And it may be as deep as 700 feet (210 m) below the surface of the water, containing a more diverse habitat than previously known.[10] The images showed sand dunes, schools of fish, black coral, sponges, rhodolith mounds, groupers, algae, and more. They also found some species that didn't seem to belong, such as a type of damselfish that normally lives in the Caribbean. Scientists believe these species show that the Amazon coral reef is the bridge between the reefs in the Caribbean and the Atlantic.

GREENPEACE

In 1971, several environmental activists sailed a fishing boat from Vancouver, Canada, to an island off the coast of Alaska. Their goal was to protest nuclear testing by the US government. And even though the US Coast Guard stopped the activists, they were able to draw worldwide attention to the dangers of nuclear testing. This was the start of Greenpeace. Today, the organization has almost three million members and offices in 55 countries.[12] Greenpeace continues to fight for many environmental issues, such as stopping the destruction of ancient forests and the deterioration of oceans.

Even with all the exploration of the Amazon reef, it is estimated that only 5 percent of it has been surveyed so far.[11] In order to more fully study the reef, it is important for it to be conserved. However, the Brazilian government has opened the area for oil development. It has awarded grants to oil companies BP, Petrobras, and Total. Greenpeace is hoping that the 2018 study, which shows the extent and importance of the Amazon coral reef, will block plans to drill in the area. However, the Amazon mouth may contain vast reserves of fossil fuels, which will attract oil companies and the Brazilian government.

CHAPTER FIVE

MEDICINE DISCOVERIES

The Amazon rain forest is home to about 80,000 plant species.[1] They make up about two-thirds of all the plants on Earth.[2] Scientists and drug companies hope to find new medicines from these plants, like how the drug quinine was discovered. Quinine, a drug that treats malaria, came from the bark of the cinchona tree in the eastern Andes.

LOOKING FOR CANCER CURES

In 2009, a floating laboratory sailed on the reddish-brown waters of the Cuieiras River. The Cuieiras is a tributary of the Rio Negro, which flows into the Amazon. On board there were researchers and scientists, including botanist Mateus Paciencia, forest guide Osmar Ferreira Barbosa, and oncologist Drauzio Varella, who was one of Brazil's most famous doctors

The Amazon rain forest is teeming with life, including undiscovered plant species that may cure human diseases.

and the founder and leader of these missions into the forest. The boat was only four hours from the city of Manaus, in Brazil's state of Amazonas, but there were no signs of other people. The thick rain forest canopy blocked out the sun, and only the yellow beaks of toucans appeared amid the greenery. Off in the distance was the white mist of a heavy rainfall. On the bank of the river stood their scientific base—little more than a wooden shack.

The group was looking for new medicine in the Amazon rain forest, a practice called bioprospecting. The researchers sought chemical compounds that could fight cancer, which they expected to find in plants and trees. In turn, they hoped their work would help save the rain forest from developers by giving people a new way of making money. Instead of logging, farming, or mining, locals could collect and sell medicinal plants. Paciencia believed this project could slow down deforestation. "You don't need to chop a single tree down to obtain these resources," he said. "You cut a little piece of the plant . . . [and] next year it will have grown back. I can't see anything more environmentally correct than a project like this."[3]

RIO NEGRO

The Rio Negro, which means "black river" in English, is a main tributary of the Amazon. It is about 1,500 miles (2,400 km) long and flows through Venezuela and Brazil.[4] The Solimões, which is full of sand, sediment, and silt from the Andes Mountains, is tan colored. In contrast, the Negro contains dissolved and decayed plant matter, which makes it black. When the two meet, they do not mix for four miles (6 km).[5] This is because of their different temperatures, densities, and speeds. The result is a dramatic swath of black water flowing next to tan water. When they eventually mix, they become part of the Lower Amazon River.

Rain forests produce about 70 percent of the known plants that have anticancer properties.[6]

The floating laboratory was from São Paulo's Paulista University. It set out every month to monitor specific plants and trees. Researchers collected samples from families of plants that had shown promise in their São Paulo laboratory, as well as samples of any new species they found. On this trip in 2009, some samples were collected by shimmying up trees and hacking down branches with a machete. After the branches crashed to the forest floor, researchers collected and sorted the samples and put them into marked bags to be studied later in the lab.

The scientists knew that every living thing—from large animals to microorganisms—is a collection of chemicals. In particular, plants are full of substances that could potentially be made into medicines. The scientists' goal was to find applications for these substances.

MEDICINE FOUND IN THE AMAZON

Many modern medicines have already been found in the Amazon rain forest. Cat's claw is a vine that indigenous healers use to treat many ailments, including stomach ulcers, fevers, and arthritis. It is being studied today as a possible cancer treatment. Cinchona is a type of tree in the Andes. Its bark has been used since 1820 to make the drug quinine to treat malaria. Jaborandi is a species of tree, also called Pilocarpus. The Guarani people have used it to heal mouth ulcers, colds, and flu since at least the 1500s. As the drug pilocarpine, it is now used in eye drops to treat high pressure in the eyes. Cordoncillo leaf is used to help stop cuts from bleeding and can keep cuts clean.

THE START OF VARELLA'S FLOATING LAB

In 1992, Varella traveled to the Amazon with Robert Gallo, a biomedical researcher from the United States. Gallo is one of two people who discovered the HIV virus. On this trip, Gallo asked Varella whether he was studying the native plants of the Amazon for medicinal

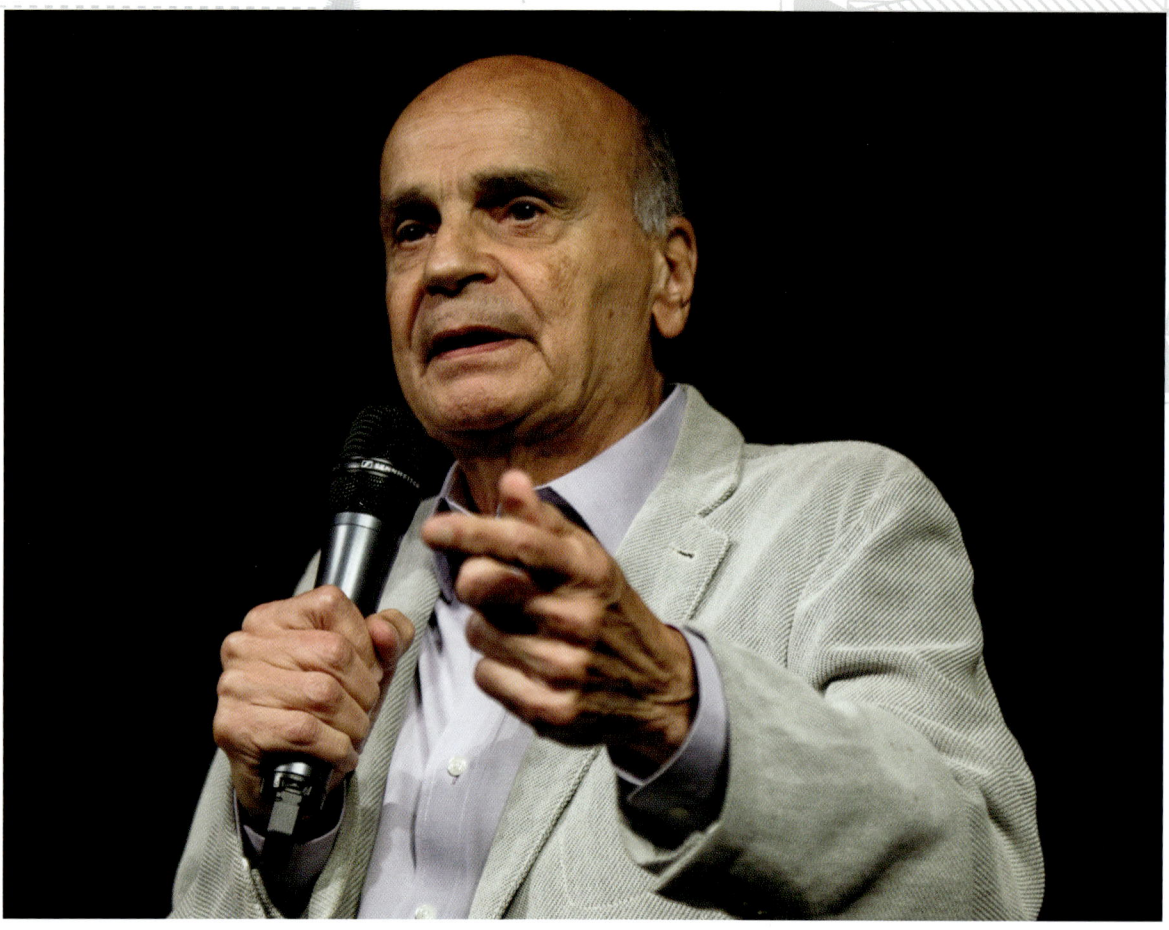

Drauzio Varella writes and speaks on medical and scientific topics.

purposes, as some other researchers were doing. Varella was not, but Gallo's question made him think that maybe he should be.

Three years later, Varella launched his first medicine-hunting expedition into the Amazon, and in 2009, he joined with São Paulo's research hospital, Sírio-Libanês. Their mission was to explore a new area of the Rio Negro, closer to Colombia.

Since then, Varella's missions have collected more than 2,200 samples. About 70 of them have shown an effect against tumors.[7] "The Amazon has something like 20% of all the biodiversity in the world," said Varella. "Just in terms of plants with flowers, there are around 22 or 23 thousand."[8] With such a rich resource, one might expect other scientists to jump at the chance to join the mission. Unfortunately, government bureaucracy and biopiracy have made it extremely difficult for pharmaceutical companies and foreign scientists to work with Varella.

Today, scientists continue to study the Amazon's trees and plants in search of medicine. It is big business for companies to bioprospect, especially in Earth's rain forests. They look for material that can be used in cosmetics, food, and medicines. Bioprospecting has become so popular that vacation companies even offer trips that include searching for medicinal plants.

BIOPIRACY

Biopiracy occurs when native materials are taken from a country without paying for them. This can include tourists stealing souvenirs or a pet store owner gathering fish for aquariums. Another kind of biopiracy involves stealing genetic material to use in researching and creating new products. The products that result—rubber, cosmetics, and medicine, to name a few—can make billions of dollars for companies. However, the native peoples and the local governments usually receive no compensation. In an attempt to combat genetic biopiracy, scientists are creating a giant database of all genetic material in the Amazon.

SHAMANS

Before scientists started looking for medicine in the Amazon rain forest, there were indigenous healers, sometimes referred to as shamans, such as Jose Roque. He is a healer from the Shipibo tribe, and he keeps a garden in the forest behind his home. He uses the

plants he grows there to treat headaches, nausea, rashes, pains, inflammation, and other ailments.

Experts go to indigenous healers for their valuable knowledge that has come from refining and experimenting with Amazon plants and animals for hundreds of generations. In the northwestern Amazon alone, it is estimated that indigenous healers use 1,300 different plants for medicinal purposes.[9]

Finding plants that can be turned into drugs is difficult. Only one out of about 10,000 to 20,000 natural compounds ever becomes medicine in the United States. Scientists want a head start, so they seek out indigenous healers. Unfortunately, being a healer is a dying art. Younger generations are distracted from learning the old traditions, and tribes are increasingly being displaced or eliminated by developers. It is estimated that about 90 tribes disappeared in the 1900s because of loggers, miners,

Indigenous elders and healers hold an incredible amount of knowledge about their forests.

"Every time a shaman dies, it is as though a book is burned."[11]

—Jose Roque, Shipibo shaman

RECORDING HEALING KNOWLEDGE

When the Matsés people of Peru and Brazil worried that their ancient medical knowledge might be lost, they created a massive encyclopedia. In 2013, five healers got together and wrote everything they knew about medicine and plants. They used their own language and included illustrations. Acaté, an organization dedicated to helping indigenous people in the Peruvian Amazon, assisted them. The first volume, 500 pages long, was released in 2015, and a second volume of similar length was released in 2017. The Matsés are using the encyclopedia to train future healers. It is the first encyclopedia of its kind to be written down.

and developers.[10] As the tribes disappeared, so did their healing knowledge.

But it's not just the plants. To understand indigenous medicine, scientists need to study and preserve the culture too. A 2009 study concluded that indigenous medicine involves examination, diagnosis, communication, ritual, and treatment. All of these aspects are important in learning which plants might work as modern medicine and how they should be used.

Leaves from the Amazon's matico plant have several medicinal uses, such as decreasing muscle fatigue.

A technician works in a drug company's product development lab.

SCIENCE CONNECTION
FROM PLANT TO DRUG

After a plant is identified as a potential drug, it undergoes rigorous and time-consuming testing. During this process, chemists and biologists work together. The chemists look at the molecules that might become drugs while the biologists look at the molecules the drugs would act on, such as cancer molecules.

The first step takes place in the lab. Scientists use robots to mix thousands of plant molecules with thousands of disease molecules. One thing they look for is to see whether one of the plant molecules will kill any disease molecules. If that happens, then variations of that plant molecule are checked to see whether a stronger version exists. These variations are also tested. After this, the molecule is used on cells and animals to imitate how the drug would work on people. Scientists want to see how effective the new drug is and whether there are any dangerous effects.

Once a drug is shown to work in animals, it is then tested on a small number of humans. This stage is called clinical trials, and its purpose is to see how effective the drug is on people and what kinds of side effects it has. If this stage goes well, then in the United States the Food and Drug Administration (FDA) approves the drug for the general public. Overall, this process takes about ten to 15 years from discovery to putting a new drug on the market.[12] It can cost pharmaceutical companies more than $1 billion to create a new drug.[13]

CHAPTER SIX

STRANGE WEATHER

A mist hangs over the Amazon rain forest, partly obscuring the treetops. It looks as though clouds have descended from the sky to sit among the trees, but these aren't clouds. They didn't come from the atmosphere. In some areas of the rain forest, the trees make the mist. Researchers suspected this mist had a major impact on the region's climate, but they didn't understand exactly how it worked.

RAINY SEASON

In 2017, scientists shed light on the phenomenon in a report published in the journal *Proceedings of the National Academy of Sciences of the United States of America* (*PNAS*). Researchers studied the southern part of the rain forest, which is a transitional zone between the tropical rain forest to the west and

The Amazon is often misty and wet.

north and the farmland and subtropical savanna to the east and south. It makes up about 30 to 40 percent of the total rain forest.[1] They looked at what caused the mist to form.

In most tropical regions, wet and dry seasons are caused by monsoons. Monsoon winds occur when the prevailing winds change direction each season. In the tropics, the prevailing winds are trade winds that blow from the east. Both the northern and southern hemispheres have their own trade winds. They converge, or meet, at the equator in something called the intertropical convergence zone (ITCZ). Another name for the ITCZ is the doldrums. Doldrum means stagnation or sluggishness, and the doldrums are an area of weak winds and calm weather.

In the doldrums, the sun heats the area, which causes air to rise and move north and south. However, the ITCZ doesn't stay in one place. Each season, it shifts in relation to the equator. This changes the air pressure and in turn creates the monsoon winds. These mainly affect Southeast Asia, Australia, and southwestern North America. But the southern part of the Amazon rain forest has monsoon winds too.

TROPICAL RAIN FOREST

Most of the Amazon rain forest lies in the tropics—an area between 23.5 degrees north and 23.5 degrees south of the equator. The Amazon River lies just south of the equator. The tropics have equal days and nights with a hot, humid, and rainy climate. Average temperatures range from 68 to 84 degrees Fahrenheit (20 to 29°C). Humidity runs about 50 percent during the day and 100 percent at night. Rainfall is at least 70 to 100 inches (180 to 250 cm) per year.[2] The water vapor in the atmosphere comes from the Atlantic Ocean in the east as well as transpiration from the forest.

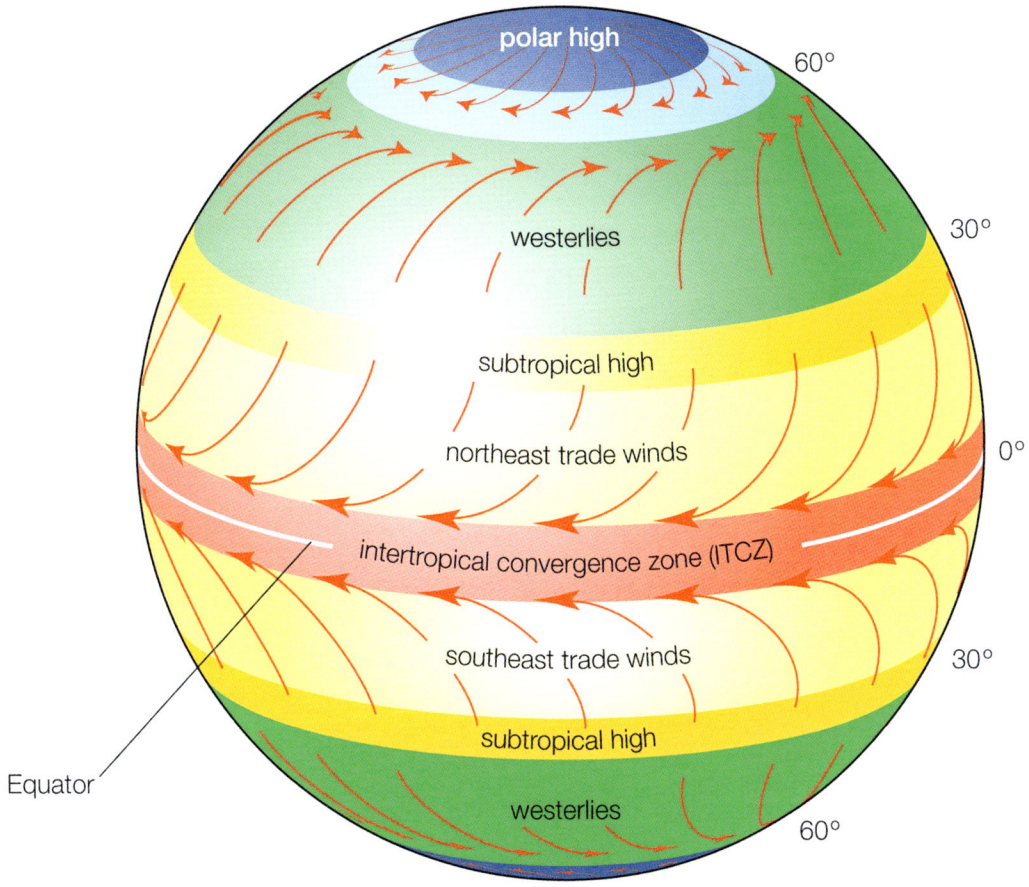

The ITCZ circles the midpoint of the planet.

The problem with the Amazon monsoon winds is that they don't bring on the wet season as they do in other parts of the world. In the Amazon, the rainy season starts in mid-October. Until now, scientists didn't have an explanation. "All you can see is the water vapor," said Rong Fu, a climate scientist at the University of California and one of the researchers on the study. "But you don't know where it comes from."[3]

CONDUCTING A STUDY

In 2004, Fu published a paper theorizing that the moisture might come from plants, specifically from transpiration. This is when plants take in water through their roots and release it through small pores in leaves called stomata. The water goes into the air as vapor. "We didn't have hard evidence," said Fu. "We speculated that the moisture came from vegetation because satellite measurements showed the vegetation became greener at the end of the dry season."[4] However, finding greener plants does not always mean more transpiration is occurring.

To prove her theory, Fu joined with other scientists to find proof of where the extra moisture was coming from. They used the Aura satellite, run by the National Aeronautics and Space Administration (NASA), to collect data on the type of water in the atmosphere over the rain forest. What they were looking for was deuterium. This is a heavy isotope of hydrogen. Normally, hydrogen atoms have just one proton and one electron. They do not have a neutron. Deuterium, however, is a hydrogen atom with a proton, an electron, and a neutron. Water with deuterium is called heavy water.

FOREST SUNLIGHT

Cloud cover limits the amount of sunshine over the Amazon rain forest. The highest part—the canopy—gets four to six hours of sun each day. But only about 2 percent of the sunlight reaches the ground.[5] This results in a dense layer of very tall trees, growing up to 120 feet (37 m).[6] The understory is more barren, full of decomposing plants and slender saplings that usually don't make it to maturity. The only way these small trees can keep growing is if there is a break in the canopy. The lucky few that get enough sun might live for centuries. However, fire is becoming more common in parts of the Amazon that have never experienced fires before, primarily as people remove trees to make cattle pastures. This fundamentally disturbs how the ecosystems behave in those regions.

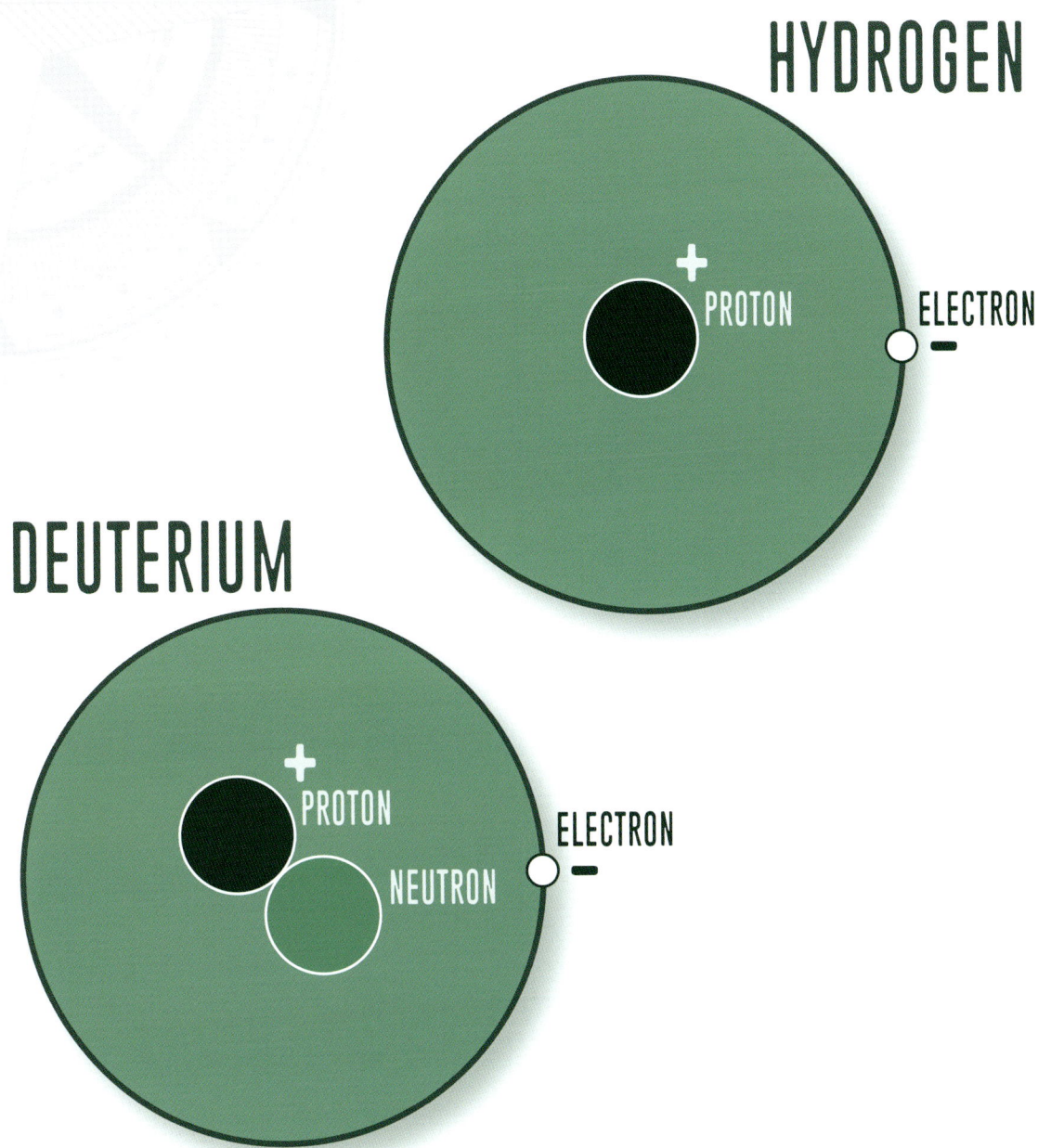

AURA

Aura, which means "breeze" in Latin and Greek, is a satellite that orbits Earth. NASA launched it in 2004 to study the atmosphere. Aura is part of a group of satellites that fly close to each other called the A-Train. Each satellite in the A-Train collects data on Earth's climate. Aura has instruments that measure the chemistry of the atmosphere, trace gases, microwave emissions from molecular vibrations, and more. It uses this data to answer questions about air quality, ozone loss, and climate change.

Deuterium was discovered in 1931 by Harold C. Urey and two colleagues. Urey won a Nobel Prize for chemistry in 1934.

When water from the ocean evaporates, the deuterium gets left behind. But when plants create water vapor through transpiration, they do not change the composition of water, and the deuterium goes up into the atmosphere. This means the water vapor from the rain forest should be different—and heavier—than the water vapor from the ocean.

John Worden from NASA's Jet Propulsion Laboratory in California, an author on the study, figured out how to distinguish between hydrogen and deuterium molecules using data from the satellite. This allowed the scientists to pinpoint where the water in the rain forest was coming from: the ocean or the plants. The results showed high levels of deuterium over the rain forest, much higher than if the water vapor had come from the ocean. In addition, Aura measured more deuterium during the greening period when plants were growing the most. This is a time when more transpiration was occurring. "What we showed," said Worden, "is that during the dry season water from vegetation is pumped into the middle troposphere where it can turn into rain."[7] The troposphere is the layer of atmosphere closest to Earth.

The study did not figure out why the plants start greening two to three months ahead of the traditional wet season. Fu speculates that the rain forest gets plenty of sun late in the dry season and that it anticipates the upcoming wet season. Thus it starts growing early. It has adapted to the extra moisture in the atmosphere. However, another study will need to be conducted to see whether this hypothesis is correct.

WHAT THE STUDY MEANS

Scientists have long researched the role forests play in weather. This study shows a connection between plants and rainfall. Transpiration may not only put additional moisture into the atmosphere but also cause the atmosphere to warm up. As the warm air rises, it circulates, possibly bringing in more moisture from the ocean.

JET PROPULSION LABORATORY

NASA's Jet Propulsion Laboratory (JPL) is a national research facility in California, but it started much more modestly. In the 1930s, several students and some friends met at the California Institute of Technology (Caltech) to work on rockets. After an explosion occurred on campus, the group moved to a remote area near the San Gabriel Mountains. In 1944, it became JPL and began working for the US Army on rocket and missile technology. Eventually JPL became part of NASA and helped initiate the exploration of space. Today, the lab works on planetary exploration, astronomy, science about Earth, and more.

The Amazon is the oldest forest in the world.

Deforestation is an issue. As the forests are cut down, the moisture stops coming from them. Stopping deforestation may help prevent or reduce droughts. Deforestation also may prevent the forests from making enough moisture to produce mists during the dry season. In the 1970s, the wet season started a month earlier than it does now. If the wet season is shortened even more, there might not be enough rainfall to keep the trees alive. "The fate of the southern Amazon rain forest depends on the length of the dry season," said Fu, "but the length of the dry season also depends on the rain forest."[8]

Increased reforestation efforts could help save the Amazon.

CHAPTER SEVEN

NEW SPECIES

The Amazon rain forest is the largest tropical rain forest in the world. Within its four million square miles (10 million sq km) live 10 percent of the world's plant and animal species.[1] Yet much of it is still undiscovered. This is due partly to its vast size and partly to a lack of money for scientific exploration.

SCIENTIFIC SURVEY

During 2014 and 2015, the World Wildlife Fund (WWF) and the Mamirauá Institute for Sustainable Development tracked all new species of plants and vertebrates found in the Amazon. Vertebrates are animals with a spine. The study did not consider insects or other invertebrates because of the vast number of those discoveries. Also, the study considered only species that that had been verified by multiple researchers. Overall, almost 400 new

The Amazon rain forest can seem to stretch on forever.

New species of insects are discovered so frequently in the Amazon it is hard to keep track.

species were discovered or described in the Amazon rain forest during the two-year period. The findings were published in 2017 in a report called "Untold Treasures: New Species Discoveries in the Amazon 2014–2015." These species included 216 plants, 93 fish, 32 amphibians, 20 mammals (including two fossils), 19 reptiles, and one bird. Previously, the WWF had surveyed 1,200 new species from 1999 to 2009, as well as 602 new species from 2010 to 2013.[2]

How could so many species remain undiscovered in the rain forest as late as 2014 and 2015? Before then, most of the samples had been taken near large towns or rivers or within protected areas. However, researchers in this survey explored regions that were more remote and difficult to get to. These included small rivers, secondary tributaries, and interfluves, which are raised areas between two rivers. In addition, deforestation may be making some areas more accessible than before, revealing species that had once been hidden.

A new species is discovered in the Amazon about every two days.[3]

FLYING MONKEY

Laura Marsh, director of the Global Conservative Institute, is an expert on flying monkeys, otherwise known as saki monkeys. She discovered five new species in 2014, but she hadn't been able to see one particular monkey species—the Vanzolini bald-faced saki. It was named after Brazilian zoologist Paulo Vanzolini and hadn't been spotted since 1956. Marsh wanted to find out whether it still existed. In the summer of 2017, Marsh and her team set out on a remote four-month expedition into the Amazon. They sailed on the Eiru River near Peru and worked with local people to search for the monkey. Finally, as the boat idled on the river, they spotted its long hair and gold-colored arms. Because it had a nonprehensile tail, one that can't be used for swinging, it ran through the trees on all fours. "It was fantastic," said Marsh about the discovery. "I was trembling and so excited I could barely take a picture."[4]

SPECIES LOSS IN THE AMAZON

Many of the Amazon's forests are clear-cut for beef production, which goes to American plates as hamburgers, and for soybeans, which increasingly feed China's pork industry. Ricardo Solar, an ecologist who has worked in the Amazon to build parks and forest reserves, is worried that habitat loss may cause tremendous loss of the native species there. Together with warmer temperatures, fires, and dam and road construction, this could trigger runaway forest degradation. That would lead to bad harvests for farmers and major problems for people living there.

Vitor Gomes, an environmental scientist at the Federal University of Pará, calculated that for trees, these forces could halve the number of species in a given area in the future. This would spell trouble for insects and other organisms that depend on those trees for survival. "For biodiversity, this could be terrible," Gomes said.[5]

RIVER DOLPHIN

One of the species discovered during the survey was a new river dolphin. This freshwater mammal is related to whales and is one of the rarest and most endangered dolphins in the world. Called boto in South America, it has a large bulge on its forehead and a long snout that it uses on the bottom of rivers to find fish and crustaceans. Botos can grow up to eight feet (2.4 m) in length and have grey, pink, or mottled coloring.

In 2013, researchers led by Tomas Hrbek of the Universidade Federal do Amazonas in Manaus discovered the new river dolphins in the Araguaia river basin. This is remarkable because other species of river dolphins live in the Amazon river basin to the west, and the two areas are almost completely isolated from each other by a series of rapids and a small canal. Just how did the Araguaian dolphins get where they were?

Researchers wondered whether they could be a new species. For 12 weeks, they went into the field and observed the dolphins. They took samples from both the Araguaian dolphins and the two other species that live in the Amazon basin, as well as samples from carcasses. After performing DNA testing, the scientists concluded that the Araguaian dolphins were different enough from other river dolphins to be classified as a new species. They estimated that the Araguaian dolphins had separated from the other dolphins more than two million years ago. This is the first new species of river dolphin discovered since 1918.

TITI MONKEYS

A new species of titi monkey was also discovered. There are more than 30 species of this monkey. They live near water sources in forests with dense growth, so they can hop from branch to branch. They are small and grow just nine to 13 inches (23 to

Researchers only recently discovered that the Amazon river dolphin has a sister species.

33 cm) long. They eat mostly fruit and live in family units of a mother, father, and offspring. The monkeys have bushy sideburns on their faces and long tails.

In 2010, Júlio César Dalponte from the Institute for the Conservation of Neotropical Carnivores and his colleagues spotted a titi monkey on the banks of the Roosevelt River. It had bright rust-red sideburns and a rust-red tail. No other titi monkeys had this coloration, so they knew it was possible they were looking at a new species. In 2010 and 2013, Dalponte led four expeditions to the Aripuanã, Roosevelt, and Guariba Rivers in Mato Grosso and Amazonas, Brazil. To collect data, they canoed on the Guariba River, walked on preexisting trails, and talked to locals. They also recorded the sounds the monkeys made and even played back the sounds to locate a group of monkeys that called back to them. They collected specimens of the new titi monkeys in nine locations as well as other titi monkeys for comparison.

In 2014, the finding was published by the Zoology Museum at the University of São Paulo in an article by Dalponte, José de Sousa da Silva Junior, and Felipe Ennes. They named the new species Milton's Titi after Brazilian professor Milton Thiago de Mello, a pioneer in primate research. "From now on, all the information we

ROOSEVELT RIVER

After losing the 1912 election, former US president Theodore Roosevelt set out on an adventure. With porters, scientists, explorers, and his adult son, Roosevelt set out to explore the River of Doubt. This tributary of the Amazon had never been explored by Europeans, and Roosevelt was warned of its potential dangers. Unfortunately, the warning was prophetic. The expedition reached the River of Doubt in February 1914 after an exhausting two-month overland trek. Things went downhill from there. The expedition faced alligators, piranhas, hostile indigenous groups, dangerous rapids, and a lack of supplies. Roosevelt almost died during the journey, but eventually they reached a relief party that was waiting for them. After this experience, Brazil renamed the river the Roosevelt River.

The Amazon is home to a huge variety of frogs, toads, and other amphibians.

have on the Milton's Titi monkey will be available to the entire scientific community," said Dalponte. "It can use the data to put together the pieces of this great jigsaw puzzle and gain a better understanding of the Amazon's biodiversity."[6]

OTHER NEW SPECIES

Researchers also found or described new fish, birds, reptiles, amphibians, and plants. These include the honeycomb stingray (*Potamotrygon limai*), named after the speckled yellowish honeycomb pattern on its dark-brown back. It was found in the Jamari River, which is in the Madeira river basin. The stingray comes in many colors and grows to 26 inches (66 cm) in length. It lives exclusively in fresh water in South America.

An example of a new bird species is Chico's tyrannulet (*Zimmerius chicomendesi*), named after Chico Mendes, a globally recognized leader for the rights of workers who tapped rubber trees. This bird likes sandy, poorly drained shrublands, grasslands, and steeper areas with stony soil that is drained well. It lives on fruit, but it also eats insects, depending on the time of year.

One of the new reptiles is the four-eyed snake (*Eutrachelophis bassleri*). It does not actually have four eyes. It has a black head and a lighter-colored pattern around its neck that resembles a necklace. From above, its patterns and colors resemble a second set of eyes. It was first discovered in 1923 by a researcher canoeing on the Pisiqui River in Peru. He reported the snake in a 1927 paper, but its discovery wasn't made official until 2014.

It can take about 20 years for a specimen to be classified as a new species.[7]

CHICO MENDES

As a boy, Chico Mendes (1944–1988) of Brazil worked alongside his father tapping rubber trees for natural rubber. When rubber prices went down, conditions for workers grew worse. Eventually the land was sold, logged, and turned into pastures for cattle ranching. In the 1970s, Mendes began fighting for workers' rights as well as for preservation of the rain forest. He pushed for forest reserves where different products could be harvested in a sustainable way. This would benefit poor communities and indigenous people. He won a Global 500 Award from the United Nations in 1987 for his activism. In 1988, he was assassinated.

One of the new amphibians in the survey is the hylid (*Tepuihyla obscura*), a type of tree toad described in 2015 and found in the Venezuelan tabletop mountains. Because this nocturnal animal is difficult to spot, the word *obscura* in its scientific name comes from the Latin word *obscurus*, which means "dark" or "hidden." These hylids live in open areas on plateau summits between 5,900 feet (1,800 m) and 8,500 feet (2,600 m) above sea level. They grow to about 1.5 inches (3.8 cm) in length.

The WWF is hoping that surveys such as this one will inform people—especially decision makers in governments—about the need for conservation in the Amazon. In turn, that will lead to more research, more preservation policies, and more protected areas. "The discovery of 381 new species is a wake-up call for the governments of Amazon countries," said Sarah Hutchison, WWF head of programming for Brazil. "They must halt the ongoing and relentless deforestation and work to preserve its unparalleled biodiversity."[8]

WWF

The WWF has a mission to conserve nature and help preserve the diversity of life on Earth. The organization began in 1961 with 16 conservationists from all over the world, representing groups such as the International Union for Conservation of Nature (IUCN) and the British Nature Conservancy. In 1961, the group knew that even though many organizations had the scientific expertise they needed to do conservation work, these organizations did not have the funding. The WWF stepped in to provide financial support for other conservation groups and for the conservation movement. Now the WWF is active in 100 countries.[9]

The Amazon Tall Tower Observatory rises about 1,070 feet (325 m) into the sky.

SCIENCE CONNECTION
FIELDWORK

Fieldwork in the Amazon involves sleeping in hammocks and getting up at the crack of dawn, then spraying on insect repellent to keep the chiggers away. It entails bumpy truck rides on narrow roads and hiking with waterproof day packs that have climbing gear, helmets, and first-aid kits. It means experiencing the rain forest firsthand, hearing the calls of macaws and the screech of monkeys and walking past termite mounds as tall as your waist, built from the iron-red earth.

And it involves the scientific work itself—going past the yellow tape that encircles the protected research area and hiking to the top of 150-foot- (46 m) tall research towers, connected to other platforms with a series of walkways. It means seeing the forest from the top of the canopy where dim light suddenly gives way to brilliant sunshine, where the trees stretch out forever like heads of broccoli, interspersed with an occasional taller tree rising above them all. And most importantly, it involves collecting data related to your study: carbon dioxide, humidity, leaf temperatures, light levels, and many other phenomena. And when the day's work is all done, it includes going back to camp and sharing a meal with fellow researchers who have come from all over the world.

Scientists who live in the Amazon can just drive to their field sites. But for others who live far away, they can't just jump on a plane and fly to the Amazon. They have to go to the appropriate governments for permits and permission to do their work.

CHAPTER EIGHT

DEFORESTATION

The Guajajara Indians motor down the muddy Caru River in the Brazilian state of Maranhão. The early evening sky is cloudy, and a wall of green lines both sides of the river. Or so it seems. On the Guajajara side of the river, the forest is untouched and pure. On the other side, just beyond a thin line of trees, the land is barren.

Today, the Guajajara are on patrol in their ancestral land, looking for illegal logging. Some of the men wear camouflage and carry weapons. Others are bare chested with dark gray paint applied to their skin.

This is a popular area for illegal logging, so there is a good chance of catching a thief in the act. And sure enough, they spot an empty canoe parked on the riverbank. The men in camouflage get off the boats and follow a narrow trail. A cut branch shows them that someone has come by recently with a machete. A moment later, they hear voices up the trail. They

Guajajara men prepare to patrol their forest for illegal logging.

Brazil has lost more Amazon rain forest than any other country.

crouch down and wait, ready to ambush whomever is coming. It turns out to be three young men who admit to cutting down trees in the Guajajara territory. They planned to make and sell charcoal. The Guajajara take the boys back to their camp for questioning.

Calling themselves "the Guardians of the Forest," the Guajajara have been patrolling the area since 2012. It's their way of protecting their land and battling illegal logging, which can lead to a decrease in wild game. Too much logging can even dry up rivers and streams. "This struggle for us is war because it is dangerous, risky," said Claudio Da Silva, their leader. "The invaders don't respect us. They want confrontation. We run into armed hunters. The loggers carry arms. The farmers are armed. It's a war in which we can die at any time."[1]

It is a war they may have to continue waging. Deforestation had been declining. It went from 11,000 square miles (28,500 sq km) per year in 2004 to 1,700 square miles (4,400 sq km) per year in 2012. However, since then, deforestation has increased once more. In 2016, deforestation was up by 29 percent.[2] In 2019, President Bolsonaro of Brazil, who prioritizes business

FOREST GUARDIANS

The roughly 120 Guardians of the Forest have made it their mission to defend their land. In so doing, they've destroyed about 200 logging sites since 2012.[3] Plus, they've taken chain saws from loggers, set trucks on fire, and chased loggers out of their territory. They also protect the Awá, one of Brazil's uncontacted tribes, who roam from place to place to avoid loggers. But the illegal logging situation isn't as simple as indigenous people versus loggers. Bruno Pereira, coordinator for uncontacted tribes at FUNAI, says that loggers have hired locals to work for them, pitting some indigenous tribes against each other.

over environmental issues, considered eliminating the National Council of the Environment. He would replace it with a handful of political appointees. Environmentalists worried this would lead to even more deforestation. It could spell an end to the lush tropical rain forest altogether.

TIPPING POINT

Two scientists, Carlos Nobre, chairman of Brazil's National Institute of Science and Technology for Climate Change, and Thomas Lovejoy, a professor at George Mason University, wrote an editorial about deforestation. It was published in 2018 in the journal *Science Advances*. They added their voices to warnings of a potential tipping point in the Amazon rain forest. A tipping point is when a series of small changes causes a much larger change to occur. In this case, the small changes are deforestation, and the larger change is the disappearance of the rain forest. If a tipping point is reached, much of the rain forest will turn into a savanna with sparse vegetation and shrubs. The enormous biodiversity currently in the rain forest will decrease immensely.

The tipping point comes because the rain forest itself makes rain, so the smaller the rain forest, the smaller the amount of rainfall. In the 1970s, Brazilian scientist Eneas Salati explored the connection between vegetation and climate. He looked at isotopic ratios of

PRESIDENT BOLSONARO

In 2018, Jair Bolsonaro was elected president of Brazil. Bolsonaro came in as an outsider, promising economic growth and security. Because Brazil was coming off its worst recession in 100 years and experiencing a large political scandal, voters were not interested in electing a career politician. After being sworn in, Bolsonaro eliminated regulations he said placed a burden on businesses. However, these changes may harm the environment and the indigenous people who live in protected areas.

oxygen in rainwater samples taken from all over the rain forest. He found that the Amazon makes about one-half of its own rainfall. Water comes from both evaporation and transpiration. This causes the lowest layer of the atmosphere to be wetter than it would be without a forest.

MODELS FOR DEFORESTATION

The first computer models of the deforestation showed the tipping point occurring at 40 percent deforestation.[4] This would cause the central, southern, and eastern regions to have less rain and a longer dry season, and the eastern area would turn into a savanna. The Andes region in the west would be less affected. This is because the mountains cause higher levels of rainfall than in other areas of the Amazon.

Indigenous women of Brazil marched in August 2019 to protest President Bolsonaro's policies.

In 2016, a report in the *Proceedings of the National Academy of Sciences* took more variables into consideration, including climate change and the use of fire to burn down forests for agriculture. With these added to the picture, the tipping point became 20 to 25 percent deforestation. Scientists don't know exactly how close the tipping point is right now.[5] But in 2018 they estimated that deforestation was at about 16 percent.[6]

In the 2010s, the Amazon was swaying back and forth from drought to flood and back again. Massive droughts occurred in 2005, 2010, and 2015–2016. In between were major floods: 2009, 2012, and 2014. Scientists believe these oscillations might be the first sign of the tipping point being reached.

Losing the lush tropical rain forest to savanna isn't the only problem that comes with deforestation. Carbon dioxide is another issue. Normally, the rain

An unusually high flood in 2009 inundated this town in the state of Maranhão in Brazil, which is down the Atlantic coast from the mouth of the Amazon River.

People around the world rally to save the Amazon and support its indigenous protectors.

forest absorbs enough carbon dioxide to account for human activity on Earth. This occurs during photosynthesis when plants take in carbon dioxide and release oxygen. Because of drought, fire, and deforestation, however, the ability of the Amazon to take in carbon dioxide has been greatly reduced. Now the carbon dioxide in the atmosphere is too high. Trees can't absorb as much of the total carbon dioxide as they used to.

Carbon dioxide in the atmosphere causes climate change, so deforestation can lead to an increase in climate change. "The effects of the increase in [carbon dioxide] in the atmosphere are going to be much higher," said Carlos Quesada, researcher from Brazil's National Institute for Amazonian Research. "And [carbon dioxide] will grow in the atmosphere [at] a much, much higher rate."[7]

PREVENTING THE TIPPING POINT

Lovejoy and Nobre have strong recommendations for preventing the rain forest from reaching the tipping point. They believe all deforestation should be completely stopped in the Amazon. In fact, more forest should be planted to give the Amazon a safety margin. They say Brazil should fulfill a pledge it made as part of the 2015 Paris Agreement on climate change and reforest 30 million acres (12 million ha) by 2030. This will reduce the deforested area to less than 20 percent. "If deforestation is brought to a full stop in the Amazon and Brazil fulfills its reforestation commitment," said Nobre, "totally deforested areas will account for approximately 16 percent–17 percent of the Amazon by 2030."[8] The hope is that such a safety margin would protect the rain forest from catastrophe.

According to the WWF, one-quarter of the Amazon rain forest will be without trees by 2030 if deforestation continues.[9]

PARIS AGREEMENT

The Paris Agreement is an international treaty. It was signed in 2015 as a response to climate change. As of January 2019, 185 countries had ratified it. The treaty aims to limit greenhouse gases, such as carbon dioxide, so that Earth's temperature goes up less than 3.6 degrees Fahrenheit (2°C) from temperatures before the industrial revolution (1700s–1800s).[10] The treaty also seeks to balance the greenhouse gases humans emit with ways to absorb those gases. For instance, carbon dioxide from gasoline-powered cars would be balanced with forests, oceans, soils, and technologies that can remove carbon dioxide from the atmosphere.

CHAPTER NINE
CLIMATE CHANGE

The Amazon rain forest helps fight against climate change. During photosynthesis, plants and trees take in carbon dioxide and give off oxygen. This makes the rain forest a carbon sink, a place where carbon dioxide is stored. Carbon dioxide is a type of greenhouse gas that is released by burning fossil fuels, such as gas, coal, and oil. The more carbon dioxide in the atmosphere, the more heat is trapped. Unfortunately, scientists believe the Amazon rain forest may be flipping from a carbon sink into a carbon source.

RESEARCHING CARBON

Global temperatures would have risen much more quickly if not for rain forests and other forests storing carbon dioxide from the atmosphere. However, temperatures are still going up. In 2019, the projection was an

The Amazon, home to a vast diversity of animals and plants, is also key to the planet's climate stability.

increase of 5.4 to 9 degrees Fahrenheit (3 to 5°C) by the end of the 2000s.[1] Such a rise could lead to more extreme weather, loss of plant and animal species, and rising sea levels. Therefore, it's important to understand exactly what is happening with carbon in the rain forest and elsewhere.

In 2017, a study was published in the journal *Science* with a question that had been asked before: Is the rain forest a carbon sink or a carbon source? The scientists were from Boston University and Woods Hole Research Center on Cape Cod. They studied satellite information from 2003 to 2014 and collected data on something different from what is usually measured. They collected carbon density information.

In the past, scientists used satellites to look at deforestation. And while these images can show major changes to the forest, they cannot show slighter degradation from things like logging, farming, invasive species, livestock grazing, gathering of wood for fuel, and more. From the air, the forest may look fine, but it could be under significant stress. "Degradation is a process where only a small portion of trees are removed from a forest," said Dr. Alessandro Baccini, lead author of the study.

MEASURING CARBON

One way to measure carbon in the Amazon is to climb a tower and take measurements from on high. Scientists have been doing so for many years. In 2009, scientists from Brazil and Germany began work on a giant tower, which they finished in 2015. Called the Amazon Tall Tower Observatory (ATTO), it is painted bright orange and rises 1,000 feet (300 m) above the forest floor. From there scientists can measure carbon dioxide, methane, and other data related to climate change. They can study the relationship between the forest and the atmosphere, as well as the soil down below.

"From a satellite image, the area will still look like an intact forest. But, when you lose even a small proportion [of] trees, you lose a significant amount of carbon."[2]

In the 2017 study, scientists gathered carbon density measurements from tropical forests all over the world. They wanted to see how the amount of carbon was changing over time. From the data, they created a computer model to look at small gains and losses of carbon over the last 12 years. The results showed that the tropical forests absorbed 480 short tons (436 metric tons) of carbon from 2003 to 2014. However, they released almost twice that—949 short tons (861 metric tons)—during the same time period. This is a net loss of 468 short tons (425 metric tons).[3] It means more carbon went into the atmosphere than was absorbed, making tropical rain forests carbon sources rather than carbon sinks.

IMPACT OF CLIMATE CHANGE ON TREES

A team of more than 100 scientists from all over the world wanted to see how climate change had been affecting trees in the Amazon over the last 30 years. They looked at records in the Amazon Forest Inventory Network (RAINFOR). The scientists discovered that trees needing the most moisture fared worse than those living in drier climates. The larger canopy trees did the best, possibly because of the higher carbon dioxide in the atmosphere. They could absorb the extra carbon dioxide and grow more than usual. In addition, pioneer trees, those that shoot up in the gaps left behind when other trees die, also fared well.

The amount of carbon emitted from the tropics is larger than earlier estimates. One reason could be that land degradation was not considered. The study found that more than two-thirds of carbon comes from degradation. As Baccini explained, "The carbon loss from land degradation is small but, because it happens a lot over a very large area, then it adds up to a lot of loss."[4]

DISAGREEMENT

Some scientists do not agree with the findings from the 2017 *Science* study. Professor Guido van der Werf from Vrije University in the Netherlands says that the 2003 to 2014 time frame is too short. Also, the study doesn't take into account small improvements to the rain forest that also increase carbon, but not in a negative way. For instance, the increase of carbon dioxide in the atmosphere would also promote forest growth.

However, experts warn that the forest will need to absorb more carbon than it does now to keep global temperatures from rising too much. To meet the goals set by the Paris Agreement, carbon dioxide will need to be extracted from the atmosphere, not just reduced.

Researchers are studying the impact of activities such as road construction that degrade the forest without removing it completely.

A biologist deploys a drone she is using to study Amazon river dolphins.

THE FUTURE

It's clear that for the Amazon rain forest to thrive, climate change and deforestation must be addressed. Deforestation means the rain forests do not absorb as much carbon dioxide as they once did, and this causes climate change to grow worse. It's important for scientists, decision makers, ecotourists, and anyone interested in the environment to help keep the rain forest alive and well. They can fight deforestation, plant new trees to reforest the land, and help reduce greenhouse gases that cause climate change. Most importantly, they can consume less.

As the largest tropical rain forest in the world, the Amazon has not been fully explored yet. Scientists need time to go in and identify new species, find vital medicines, help indigenous communities, explore remote areas, and conduct experiments on land, water, and air. They want to experience the thrill of being the first to discover something new—perhaps something that can be a benefit to all of us.

The average temperature on Earth has risen two degrees Fahrenheit (1°C) since the 1880s, and the most recent years have been the hottest ever recorded.[5]

EXTRACTING CARBON

Aside from using plants and trees, how else can carbon be extracted from the atmosphere? In 2018, scientists from almost 200 countries met in Poland to discuss answers to this question.[6] So far, no one has come up with an economical solution, but researchers are working on it. Carbon Engineering in Canada has come up with a direct-capture facility that pulls air into a chemical solution, which extracts and keeps carbon dioxide. Since air contains so little carbon dioxide, about 0.04 percent, the solution is heated, chemicals are added, and it is heated again to make a carbon dioxide–rich gas. The final product could be put underground or turned into something else, such as fuel. Climeworks, a company in Switzerland, is doing the same work in Europe. Both are working to make the technology worth the cost.

ESSENTIAL FACTS

SIGNIFICANT EVENTS

- In 2018, a study revealed ancient earthworks in the southern rim of the Amazon. These provide evidence about the large population of pre-Columbian people who lived in the Amazon.

- The first photographic evidence of the coral reef at the mouth of the Amazon was obtained in 2017 and 2018. These films revealed a reef up to six times larger than originally thought. They also showed that the reef extends as deep as 720 feet (220 m) below the surface of the ocean.

- The World Wildlife Fund and Mamirauá Institute for Sustainable Development surveyed all new species discovered in the rain forest during a two-year period (2014–2015). They tallied almost 400 new species, a new species every other day.

KEY PLAYERS

- Sydney Possuelo is responsible for creating the National Indian Foundation (FUNAI) and setting aside protected land for indigenous tribes.

- Jonas Gregorio de Souza is an archaeologist at the University of Exeter. He was first author of a study that determined the extent of pre-Columbian people in the Amazon.

- Drauzio Varella is an oncologist who leads missions into the rain forest in search of cancer medicine.

- Rong Fu is a climate scientist at the University of California who helped determine that rain forest plants in the Amazon are creating their own water vapor, which leads to an early wet season.

IMPACT ON SCIENCE

The Amazon rain forest is the largest in the world, with more species than any other ecosystem on Earth. By studying its plants, trees, animals, and more, researchers gain a greater understanding of medicine, the rain forest, and climate change that can lead to more advances in science.

QUOTE

"The Amazon has something like 20% of all the biodiversity in the world. Just in terms of plants with flowers, there are around 22 or 23 thousand."

—*Drauzio Varella*

GLOSSARY

botanist
A scientist who studies plants.

continental shelf
A shallow underwater plain of varying width that borders a continent and ends with a steep drop to the ocean floor.

deforestation
The action of clearing a large group of trees.

ecosystem
A community of interacting organisms and their environment.

floodplain
Land adjacent to a river that is subject to flooding.

indigenous
Originating in or native to a place.

isotope
A form of an element with a differing number of neutrons in its nucleus.

oncologist
A doctor who diagnoses and treats cancer.

plume
The place where river water flows into the ocean.

river basin
The land drained by a river and its tributaries.

savanna
A treeless plain.

sonar
A device for detecting or locating objects, especially underwater, by using sound waves that are reflected by the objects.

trade winds
Winds that blow almost continually toward the equator from the northeast and from the southeast.

transpiration
The passage of watery vapor from a plant or living body through a membrane or pores.

tributary
A stream or river feeding a larger stream, river, or lake.

variable
A factor in a scientific experiment that may be subject to change.

ADDITIONAL RESOURCES

SELECTED BIBLIOGRAPHY

"Amazon Forest Guardians Fight to Prevent Catastrophic Tipping Point." *PBS*, 13 Sept. 2018, pbs.org. Accessed 14 Aug. 2019.

Jung, Elaine. "Amazon Discoveries," *BBC*, 8 Apr. 2018, bbc.co.uk. Accessed 14 Aug. 2019.

"Rain Forests," *National Geographic*, n.d., nationalgeographic.com. Accessed 14 Aug. 2019.

FURTHER READINGS

Hand, Carol. *Bringing Back Our Tropical Forests*. Abdo, 2018.

Harris, Duchess, with Rebecca Rowell. *The Paris Climate Agreement*. Abdo, 2019.

Jackson, Tom. *The Amazon*. DK, 2015.

ONLINE RESOURCES

To learn more about Amazon explorers, please visit **abdobooklinks.com** or scan this QR code. These links are routinely monitored and updated to provide the most current information available.

MORE INFORMATION

For more information on this subject, contact or visit the following organizations:

THE NATURE CONSERVANCY

4245 N. Fairfax Dr., Suite 100

Arlington, VA 22203

703-841-5300

nature.org

This organization works to conserve lands and waters throughout the world.

RAINFOREST FOUNDATION US

1000 Dean St., Suite 430

Brooklyn, NY 11238

212-431-9098

rainforestfoundation.org

This organization is working to protect the rain forests of Central and South America.

SURVIVAL INTERNATIONAL

PO Box 26345

San Francisco, CA 94126

510-858-3950

survivalinternational.org

This organization is dedicated to protecting indigenous peoples and their lands worldwide.

SOURCE NOTES

CHAPTER 1. UNCONTACTED PEOPLE

1. "Amazon Tribes." *Survival International*, n.d., survivalinternational.org. Accessed 4 Sept. 2019.

2. Sean Kane. "More Than 100 Tribes Across the World Still Live in Total Isolation from Society." *Independent*, 8 Mar. 2018, independent.co.uk. Accessed 4 Sept. 2019.

3. "From the Boa to the Leafcutter Ant, and Back to the Red Piranha, Amazon Wildlife Comes in All Shapes and Sizes." *WWF*, n.d., wwf.panda.org. Accessed 4 Sept. 2019.

4. Rhett A. Butler. "10 Facts about the Amazon Rainforest." *Mongabay*, 1 Apr. 2019, rainforests.mongabay.com. Accessed 4 Sept. 2019.

5. Rhett A. Butler. "The Amazon Rainforest: The World's Largest Rainforest." *Mongabay*, 1 Apr. 2019, rainforests.mongabay.com. Accessed 4 Sept. 2019.

6. Scott Wallace. *The Unconquered*. Crown/Archetype, 2011. 254.

7. Rachel Nuwer. "Anthropology: The Sad Truth about Uncontacted Tribes." *BBC*, 4 Aug. 2014, bbc.com. Accessed 4 Sept. 2019.

CHAPTER 2. MYSTERIOUS EARTHWORKS

1. Jonas Gergorio de Souza et al. "Pre-Columbian Earth-Builders Settled along the Entire Southern Rim of the Amazon." *Nature Communications*, 27 Mar. 2018, nature.com. Accessed 4 Sept. 2019.

2. Nicola Davis. "Lost Amazon Villages Uncovered by Archaeologists." *Guardian*, 27 Mar. 2018, theguardian.com. Accessed 4 Sept. 2019.

3. Davis, "Lost Amazon Villages Uncovered by Archaeologists."

4. Davis, "Lost Amazon Villages Uncovered by Archaeologists."

5. "Parts of the Amazon Thought Uninhabited Were Actually Home to up to a Million People." *University of Exeter*, 27 Mar. 2018, exeter.ac.uk. Accessed 4 Sept. 2019.

6. Davis, "Lost Amazon Villages Uncovered by Archaeologists."

7. Ben Panko. "The Supposedly Pristine, Untouched Amazon Rainforest Was Actually Shaped by Humans." *Smithsonian*, 30 Mar. 2017, smithsonianmag.com. Accessed 4 Sept. 2019.

8. Alexander Koch et al. "European Colonization of the Americas Killed 10 Percent of World Population and Caused Global Cooling." *PRI*, 31 Jan. 2019, pri.org. Accessed 4 Sept. 2019.

9. C. Levis et al. "Persistent Effects of Pre-Columbian Plant Domestication on Amazonian Forest Composition." *Science*, 3 Mar. 2017, science.sciencemag.org. Accessed 4 Sept. 2019.

10. Robinson Meyer. "The Amazon Rain Forest Was Profoundly Changed by Ancient Humans." *Atlantic*, 2 Mar. 2017, theatlantic.com. Accessed 4 Sept. 2019.

11. Meyer, "The Amazon Rain Forest Was Profoundly Changed."

CHAPTER 3. BRAZIL'S HIGHEST MOUNTAIN

1. Dom Phillips. "Illegal Mining in Amazon Rain Forest Has Become an 'Epidemic.'" *Guardian*, 10 Dec. 2018, theguardian.com. Accessed 4 Sept. 2019.

2. Elaine Jung. "The 9 Unknown Species That a Group of Scientists Discovered When Entering the Pico da Neblina Mountain, a Remote Area of the Amazon." *BBC News Mundo*, 6 Apr. 2018, bbc.com. Google translation from Spanish. Accessed 4 Sept. 2019.

3. Jung, "The 9 Unknown Species."

4. Karina Toledo. "Expeditions to Amazonia Reveal New Species of Toads, Lizards, Birds and Plants." *Agencia FAPESP*, 18 July 2018, agencia.fapesp.br. Accessed 4 Sept. 2019.

5. "Yanomami." *Encyclopedia Britannica*, 2 May 2019, britannica.com. Accessed 4 Sept. 2019.

6. "The Yanomami." *Survival International*, n.d., survivalinternational.org. Accessed 4 Sept. 2019.

7. João Fellet. "Newly Discovered Species at Fog Peak Already at Risk of Being 'Strangled.'" *BBC News Brasil*, 28 Apr. 2018, bbc.com. Google translation from Portuguese. Accessed 4 Sept. 2019.

8. Elaine Jung. "Amazon Discoveries." *BBC*, 6 Apr. 2018, bbc.co.uk. Accessed 4 Sept. 2019.

9. Amber Beavis. "Name This Creature: How to Scientifically Name a Species." *ABC*, 4 Aug. 2016, abc.net.au. Accessed 4 Sept. 2019.

10. Toledo, "Expeditions to Amazonia Reveal New Species."

CHAPTER 4. THE AMAZON'S CORAL REEF

1. Amelia Urry. "New Discoveries about Unique Amazon Coral Reef at Risk from Oil Drilling." *Oceans Deeply*, 8 May 2018, newsdeeply.com. Accessed 4 Sept. 2019.

2. Robinson Meyer. "Scientists Have Discovered a 600-Mile Coral Reef." *Atlantic*, 21 Apr. 2016, theatlantic.com. Accessed 4 Sept. 2019.

3. Meyer, "Scientists Have Discovered a 600-Mile Coral Reef."

4. Jareen Imam. "Massive Coral Reef Discovered in the Amazon." *CNN*, 26 Apr. 2016, cnn.com Accessed 4 Sept. 2019.

5. Craig Welch. "Surprising, Vibrant Reef Discovered in the Muddy Amazon." *National Geographic*, 22 Apr. 2016, nationalgeographic.com. Accessed 4 Sept. 2019.

6. Rachel Nuwer. "Shining Light on Brazil's Secret Coral Reef." *Smithsonian*, 22 Apr. 2016. smithsonianmag.com. Accessed 4 Sept. 2019.

7. Welch, "Surprising, Vibrant Reef Discovered in the Muddy Amazon."

8. "What Is a Sponge?" *NOAA National Ocean Service*, 14 Mar. 2019, oceanservice.noaa.gov. Accessed 4 Sept. 2019.

9. "DeepWorker." *NOAA Ocean Explorer*, 16 Apr. 2013, oceanexplorer.noaa.gov. Accessed 4 Sept. 2019.

10. Urry, "New Discoveries about Unique Amazon Coral Reef."

11. Damian Carrington. "First Images of Unique Brazilian Coral Reef at Mouth of Amazon." *Guardian*, 31 Jan. 2017, theguardian.com. Accessed 4 Sept. 2019.

12. "About." *Greenpeace*, n.d., greenpeace.org. Accessed 4 Sept. 2019.

CHAPTER 5. MEDICINE DISCOVERIES

1. "The Giant Kapok Tree, the Creeping Aroids, and Other Resident Architects of the Amazon Rainforest." *WWF*, n.d., wwf.panda.org. Accessed 4 Sept. 2019.

2. "Rainforests, Explained." *National Geographic*, n.d., nationalgeographic.com. Accessed 4 Sept. 2019.

SOURCE NOTES CONTINUED

3. Tom Phillips. "Brazilian Explorers Search 'Medicine Factory' to Save Lives and Rainforest." *Guardian*, 27 Apr. 2009, theguardian.com. Accessed 4 Sept. 2019.

4. "Negro." *Amazon Waters*, n.d., amazonwaters.org. Accessed 4 Sept. 2019.

5. "What Causes Brazil's 'Meeting of the Waters'?" *Science Alert*, 25 June 2014, sciencealert.com. Accessed 4 Sept. 2019.

6. "Rainforests, Explained."

7. Stuart Grudgings. "In Amazon, a Frustrated Search for Cancer Cures." *Reuters*, 16 Nov. 2009, reuters.com. Accessed 4 Sept. 2019.

8. Phillips, "Brazilian Explorers Search 'Medicine Factory.'"

9. Simeon Tegel. "Dying Amazon Healers Are Taking Potential Cures for Cancer, AIDS and Other Diseases With Them." *PRI*, 29 June 2015, pri.org. Accessed 4 Sept. 2019.

10. Tegel, "Dying Amazon Healers."

11. Tegel, "Dying Amazon Healers."

12. Phillips, "Brazilian Explorers Search 'Medicine Factory.'"

13. "Biopharmaceutical Research and Development: The Process behind New Medicines." *Pharmaceutical Research and Manufacturers of America*, 2015, phrma.org. Accessed 4 Sept. 2019.

CHAPTER 6. STRANGE WEATHER

1. Jonathan S. Wright et al. "Rainforest-Initiated Wet Season Onset Over the Southern Amazon." *PNAS*, 20 July 2017, pnas.org. Accessed 4 Sept. 2019.

2. Jeremy M. B. Smith. "Tropical Rainforest." *Encyclopedia Britannica*, 24 Aug. 2018, britannica.com. Accessed 4 Sept. 2019.

3. Ilima Loomis. "Trees in the Amazon Make Their Own Rain." *Science*, 4 Aug. 2017, sciencemag.org. Accessed 4 Sept. 2019.

4. Carol Rasmussen. "New Study Shows the Amazon Makes Its Own Rainy Season." *Jet Propulsion Laboratory*, 17 July 2017, jpl.nasa.gov. Accessed 4 Sept. 2019.

5. "Rain Forest." *UXL Encyclopedia of Science*. Gale, 2002. *Encyclopedia.com*. Accessed 4 Sept. 2019.

6. Smith, "Tropical Rainforest."

7. Rasmussen, "New Study Shows the Amazon Makes Its Own Rainy Season."

8. Rasmussen, "New Study Shows the Amazon Makes Its Own Rainy Season."

CHAPTER 7. NEW SPECIES

1. Sarah Gibbens. "New Amazon Species Discovered Every Other Day." *National Geographic*, 2 Sept. 2017, nationalgeographic.com. Accessed 4 Sept. 2019.

2. "Untold Treasures: New Species Discoveries in the Amazon 2014–2015." *WWF*, 2017, wwf.org.uk. Accessed 4 Sept. 2019.

3. Jorge Eduardo Dantas. "Monkey Species Discovered by WWF-Brazil Expedition Receives Scientific Description." *WWF*, 23 Feb. 2015, wwf.org.br. Accessed 4 Sept. 2019.

4. "381 New Species Discovered in the Amazon." *WWF*, 5 Sept. 2017, wwf.org.uk. Accessed 4 Sept. 2019.

5. "Untold Treasures."

6. Nic Fleming. "How to Discover a Brand-New Lifeform." *BBC Earth*, 16 Jan. 2015, bbc.com. Accessed 4 Sept. 2019.

7. Sarah Gibbens. "Mysterious Amazon Animal Seen Alive for First Time in 80 Years." *National Geographic*, 25 Aug. 2017, nationalgeographic.com. Accessed 4 Sept. 2019.

8. Yessenia Funes. "Deforestation and Climate Change Could Split the Amazon Rainforest in Two, Study Finds." *Gizmodo*, 25 June 2019, earther.gizmodo.com. Accessed 4 Sept. 2019.

9. "Overview." *WWF*, n.d., worldwildlife.org. Accessed 4 Sept. 2019.

CHAPTER 8. DEFORESTATION

1. "Amazon Forest Guardians Fight to Prevent Catastrophic Tipping Point." *PBS News Hour*, 13 Sept. 2018, pbs.org. Accessed 4 Sept. 2019.

2. Philip Fearnside. "Business as Usual: A Resurgence of Deforestation in the Brazilian Amazon." *Yale Environment 360*, 18 Apr. 2017, e360.yale.edu. Accessed 4 Sept. 2019.

3. Karia Mendes. "Fears over Rising Violence in Amazon as 'Forest Guardians' Battle Logging." *Reuters*, 13 May 2019, reuters.com. Accessed 4 Sept. 2019.

4. Thomas E. Lovejoy and Carlos Nobre. "Amazon Tipping Point." *Science Advances*, 21 Feb. 2018, advances.sciencemag.org. Accessed 4 Sept. 2019.

5. FAPESP. "Amazon Deforestation Is Close to Tipping Point." *Phys.org*, 20 Mar. 2018, phys.org. Accessed 4 Sept. 2019.

6. Sam Eaton. "Tropical Forests Are Flipping from Storing Carbon to Releasing It." *Nation*, 30 Aug. 2018, thenation.com. Accessed 4 Sept. 2019.

7. "Amazon Forest Guardians Fight to Prevent Catastrophic Tipping Point."

8. FAPESP, "Amazon Deforestation Is Close to Tipping Point."

9. Mendes, "Fears over Rising Violence in Amazon."

10. "Paris Agreement." *Encyclopedia Britannica*, 8 Apr. 2019, britannica.com. Accessed 4 Sept. 2019.

CHAPTER 9. CLIMATE CHANGE

1. "Global Temperatures on Track for 3–5 Degree Rise by 2100: U.N." *Reuters*, 29 Nov. 2018, reuters.com. Accessed 4 Sept. 2019.

2. Daisy Dunne. "Tropical Forests Are No Longer Carbon Sinks Because of Human Activity." *CarbonBrief*, 18 Sept. 2017, carbonbrief.org. Accessed 4 Sept. 2019.

3. Dunne, "Tropical Forests Are No Longer Carbon Sinks."

4. Dunne, "Tropical Forests Are No Longer Carbon Sinks."

5. "2018 Fourth Warmest Year in Continued Warming Trend, According to NASA, NOAA." *NASA Global Climate Change*, 6 Feb. 2019, climate.nasa.gov. Accessed 4 Sept. 2019.

6. Jeff Brady. "How One Company Pulls Carbon from the Air, Aiming to Avert a Climate Catastrophe." *NPR*, 10 Dec. 2018, npr.org. Accessed 4 Sept. 2019.

INDEX

Agriculture Ministry (Brazil), 13
amphibians, 33–37, 39, 73, 77–79
Andes region, 6, 48, 50, 51, 87

Baccini, Alessandro, 94–95
Barbosa, Osmar Ferreira, 48
biodiversity, 6, 8, 48, 53, 70, 74, 77, 79, 85
biopiracy, 53
bioprospecting, 50–51, 53
birds, 6, 9, 37, 73, 77–78
Bolsonaro, Jair, 13–14, 84, 85

carbon capture, 99
carbon dating, 21
carbon density measurements, 94–95
carbon dioxide, 23, 42, 80, 88–91, 92–99
carbon sink, 92–95
climate change, 30, 34, 43, 66, 88–91, 92–99
cloud cover, 15, 19, 32, 60, 64
computer modeling, 22, 28, 34, 87, 95
coral reef, 40–47
curare, 9

Da Silva, Claudio, 84
Dalponte, Júlio Césár, 76–77
de Souza, Jonas Gregorio, 19, 22
DeepWorker submersible, 45
deforestation, 8, 19, 50, 68, 73, 79, 82–91, 94, 99
deuterium, 64–66
development in the Amazon, 13–14, 50, 55–56
    food production, 53, 64, 74, 78
    logging, 4, 6, 8, 11, 50, 55, 78, 82–84, 94
    mining, 4, 6, 8, 32, 50, 55
    oil drilling, 11, 47
drought, 68, 88, 90
drug development, 48, 51, 53, 55, 58

equator, 6, 33, 62
European conquest, 16, 23

fieldwork, 80
fire damage, 64, 74, 88, 90
Flecheiros, 4–6, 8–10
floodplains, 6, 21
floods, 6, 88
fossil fuels, 47, 92
Fu, Rong, 63–64, 67–68

Gallo, Robert, 51–52
geoglyphs, 18–19
Gomes, Vitor, 74
Greenpeace, 45–47
Guajajara Indians, 82–84
Guardians of the Forest, 84

Hill, Kim, 11

indigenous healers, 51, 53–56
indigenous tribes, 4, 8, 34
    and anthropologists, 10–11
    and disease, 6–7, 11, 23
    and forest protection, 13–14, 32, 82–84
    and technology, 14, 15
International Code of Zoological Nomenclature, 37
intertropical convergence zone (ITCZ), 62

land degradation, 74, 94–95
Levis, Carolina, 24–25
lizards, 34–35, 39
Lovejoy, Thomas, 85, 91

Mamirauá Institute for Sustainable Development, 70
Marsh, Laura, 73
Matsés people, 56
medicinal plants, 34, 48–56, 99
Mendes, Chico, 78
monkeys, 9, 34, 73, 75–77, 80
monsoons, 62–63
Moura, Rodrigo, 42–43

National Indian Foundation (FUNAI), 8, 13–14, 84
Nobre, Carlos, 85, 91

observatory towers, 15, 80, 94

Paciencia, Mateus, 48–50
Paiter-Surui tribe, 14, 15
Paris Agreement, 91, 97
photosynthesis, 23, 42, 43, 90, 92
Pico da Neblina, 30–37
Possuelo, Sydney, 6, 8–9, 14
Prates, Ivan, 35, 37
pre-Columbian societies
   earthworks, 18–22, 26, 28
   plant domestication, 23–26
   population, 16, 22–23

quinine, 48, 51

reforestation, 23, 91, 99
river dolphins, 74–75
river plume, 40–43
Rodrigues, Miguel Trefaut, 33
Roosevelt, Theodore, 76
Roque, Jose, 53–55, 56

Salati, Eneas, 85–87
satellites, 15, 19–21, 64–66, 94–95
savanna, 6, 18, 33, 62, 85–88
sediment, 40, 42, 50
Shipibo tribe, 53, 56
snakes, 78
sonar, 42, 45
species discovery, 30–39, 70–79
species loss, 74, 94
sponges, 42–44, 45, 47
Steege, Hans ter, 23
stingrays, 77–78
Surui, Almir Narayamoga, 15

terra preta (dark soil), 22, 26
tipping point, 85–91
tourism, 11, 32, 53, 99
transpiration, 62, 64–67, 87

uncontacted tribes, 4, 8, 11, 14–15, 84
Uru-Eu-Wau-Wau tribe, 13–14
van der Werf, Guido, 97
Varella, Drauzio, 48–53
Worden, John, 66
World Wildlife Fund (WWF), 70–73, 79, 91

Yager, Patricia, 42, 44
Yanomami people, 32–33, 34

ABOUT THE AUTHOR

Andrea Pelleschi has worked as an editor in the education market for more than 15 years. She has also written many nonfiction books for children and young adults.